CW01431961

ALSO BY SIMON WINCHESTER

Knowing What We Know: The Transmission of Knowledge: From Ancient Wisdom to Modern Magic

Land: How the Hunger for Ownership Shaped the Modern World

In Holy Terror: Reporting the Ulster Troubles

American Heartbeat: Notes from a Midwestern Journey

Their Noble Lordships: Class and Power in Modern Britain

Stones of Empire: The Buildings of the Raj (photographer)

Outposts: Journeys to the Surviving Relics of the British Empire

Prison Diary, Argentina: A Falklands Story

Hong Kong: Here Be Dragons

Korea: A Walk Through the Land of Miracles

Pacific Rising: The Emergence of a New World Culture

Small World

Pacific Nightmare: How Japan Starts World War III (fiction)

The River at the Center of the World: A Journey Up the Yangtze, and Back in Chinese Time

The Professor and the Madman: A Tale of Murder, Insanity, and the Making of the Oxford English Dictionary

The Fracture Zone: My Return to the Balkans

The Map That Changed the World: William Smith and the Birth of Modern Geology

THE Breath OF THE Gods

Krakatoa: The Day the World Exploded

The Meaning of Everything: The Story of the Oxford English Dictionary

A Crack in the Edge of the World: America and the Great California Earthquake of 1906

The Man Who Loved China: The Fantastic Story of the Eccentric Scientist Who Unlocked the Mysteries of the Middle Kingdom

Atlantic: Great Sea Battles, Heroic Discoveries, Titanic Storms, and a Vast Ocean of a Million Stories

Pacific: Silicon Chips and Surfboards, Coral Reefs and Atom Bombs, Brutal Dictators and Fading Empires

West Coast: Bering to Baja

East Coast: Arctic to Tropic

Skulls: An Exploration of Alan Dudley's Curious Collection

The Alice Behind Wonderland

The Men Who United the States: America's Explorers, Inventors, Eccentrics, and Mavericks, and the Creation of One Nation, Indivisible

When the Earth Shakes: Earthquakes, Volcanoes, and Tsunamis

When the Sky Breaks: Hurricanes, Tornadoes, and the Worst Weather in the World

The Perfectionists: How Precision Engineers Created the Modern World

The End of the River

Mississippi River: Headwaters and Heartland to Delta and Gulf

The Man with the Electrified Brain

THE Breath OF THE Gods

THE HISTORY AND FUTURE OF THE WIND

Simon Winchester

WILLIAM
COLLINS

William Collins
An imprint of HarperCollins*Publishers*
1 London Bridge Street
London SE1 9GF

WilliamCollinsBooks.com

HarperCollins*Publishers*
Macken House, 39/40 Mayor Street Upper,
Dublin 1, D01 C9W8, Ireland

First published in Great Britain in 2025 by William Collins
First published in the US in 2025 by Harper

1

Designed by Elina Cohen
Ship art © clu / Getty Images

A catalogue record for this book is available from the British Library

ISBN 978-0-00-867949-1 (Hardback)
ISBN 978-0-00-867950-7 (Trade paperback)

Set in Bell MT Std

Printed and bound in the UK using 100% renewable electricity at CPI Group (UK) Ltd

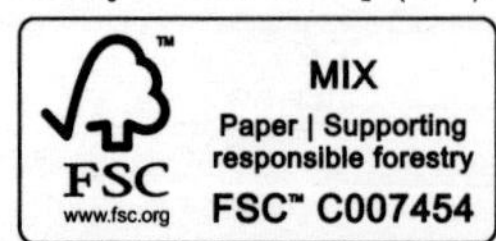

I dedicate this book, with inestimable thanks

To my friend

Thomas Dyja

Who, being a native of Chicago

Knows a thing or two

About Wind

But I can see where the wind goes
And follow the way of the wind;
And blessedness goes where the wind goes
And when it is gone we are dead.

—from "The Blessed," in *The Wind Among the Reeds,*
W. B. Yeats, 1899

CONTENTS

ILLUSTRATIONS

PRELUDE

An Unexpected Occurrence at "the Windiest Place in the World"

The whaleback summit of Mount Washington, dominant among the White Mountains of central New Hampshire, can be a truly tempestuous place, magnificent in the ferocity of its weather. It is said, with much statistical justification, to be the windiest place in the world. Until very recently it held the record for the highest measured wind speed on the planet—231 miles per hour—until an Australian island recorded a cyclonic gust that was ten miles per hour faster.

During the winters, the lightly compensated observers whose task it is—and has been for more than a century—to note the wind, the temperature, the atmospheric pressure, and the humidity every hour of every day and every night do battle with truly terrifying phenomena. Protected as best they can be with a multitude of insulating layers, they use enormous rubber-headed mallets to smash away thick accumulations of ice and blown snow to reach and try to read their instruments. Spare a thought one coming midnight for the duty observer having to wrench open the door to the howling furies outside and clamber up a freezing

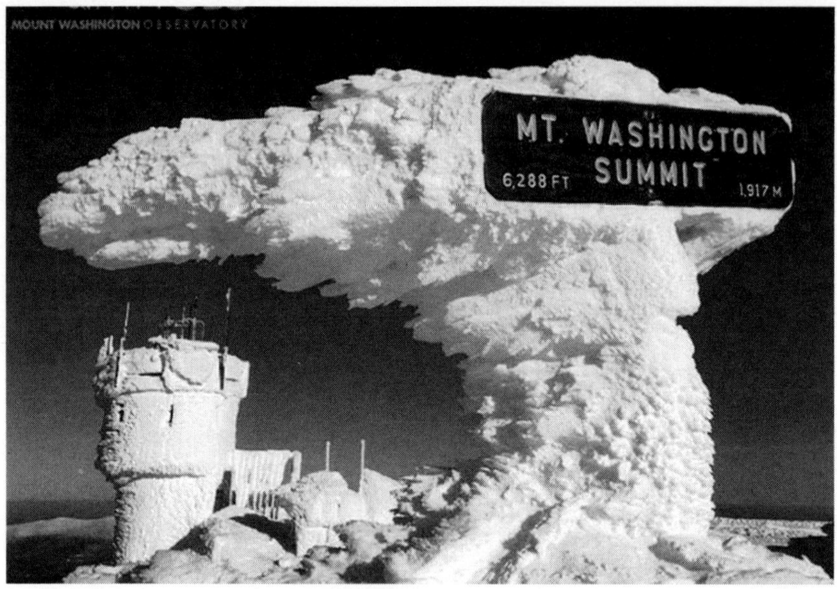

Ferocious winds on New Hampshire's highest peak carve snow into fantastical shapes, challenging those resident observers who, from the drum-shaped structure behind, must take hourly readings, night and day.

iron stairway and a vertical ladder to the observation deck, there to acquire, each hour, the figures for the statistical manuals—to enhance and complete the data sets—from which others in warm offices far away may gain the material for forecasts, for analysis, to discern trends, to gather hints as to just how the world's wind patterns might be altering, as they are said to be. Unsung heroes they surely are; spare a thought indeed.

THERE ARE DAYS up on Mount Washington, however, that are anything but ferocious. I was there on September 28, 2024, and this was one such day. A rarity. A climatic outlier. And as you shall see, in the context of the book you are holding, something of an irony.

• • •

IT WAS A perfect American fall day. From down in the Ammonoosuc River valley the great granite peaks of the Presidential Range—Mounts Washington, Jefferson, Madison, Monroe, Pierce, Adams, Quincy Adams, Eisenhower, and, though not yet fully and federally recognized, Reagan—positively gleamed against a cloudless sapphire sky. From time to time a fine mist seemed to settle on the higher peaks, but it was soon burned off by the warming sun. The leaves on the lower slopes were already turning a brilliant yellow; in another few days they would change to flaming orange, then deep red and finally purple, the endless litany of the autumn beauty to which thousands are attracted each season.

To reach the 6,288-foot summit we took the old cog railway, an ancient funicular built in 1868 and little changed since: one of the small locomotives is steam-powered, and gouts of black coal smoke rising into the sky provide reassurance to passersby that The Cog, as it is locally known, is still in fine working fettle.

Up top, a thousand feet above the tree line and often above the clouds, stands the weather station—a drum-shaped structure of brick and stone, topped with an observation floor that is itself surmounted by an instrument tower with radio aerials and microwave dishes. Nearby is a very much taller gantry festooned with yet more dishes and signaling apparatus. This, it is somewhat quietly explained, is mainly for military and intelligence purposes—the Pentagon, the CIA, and the NSA all have an electronic presence on the summit, and on the afternoon of the 9/11 attacks in 2001 a Humvee suddenly appeared with armed soldiers aboard, checking on what is regarded as a vital component of America's critical security infrastructure.

Nothing critical in the weather room, however. All here was quiet, serious, somewhat studious. Two young observers were on duty, Alexandra and Charlie, together with a pale gray weather cat named Nimbus who sat among the instruments or else paced, tail up, sliding silkily around the visitors.

A large screen on the north wall showed the current conditions—the barometric pressure, the relative humidity, the dew point, the temperature, the time, the wind speed and its direction. In the center of the screen was a large dial, like the speedometer of a car. It was marked with mph at the top, and it had a scale reading from zero up to 100. This afternoon the needle was down hard to the left, at zero, as in a car with the ignition on but that has not yet started to move. At first I thought little of it.

By now Alexandra was busily showing us charts and diagrams to explain Mount Washington's unique position in America's meteorological landscape. One chart showed historical data of storm tracks making their way eastward from the mountainous far west—thick red lines crossing the country from Oregon and California and Arizona and the High Sierra and the deserts of the Great Basin, lines showing their aggregated history as they sped south of the Great Lakes, converging steadily until most were packed together almost as one as they passed right over the New England states, then reached the sudden high mountainous barrier of this, our Presidential Range, and tried to squeeze past the peaks on their way to the sea.

But here, Alexandra explained, the Venturi effect kicks in, comes into play, the gales and storms now being compressed between two peaks into a tight corner, and then, as you may recall from school, the wind speed increases and the temperature falls—and hence Mount Washington, with the worst weather in the world.

We went over for a more detailed look at the screen. The red pointer on the large round speedometer dial was still bumped down against its stop, at zero. Otherwise, the pixels displayed an endless train of numbers—integers, whole numbers and fractions, percentages of numbers, directions in degrees, temperatures in centigrade and Celsius and Kelvin and Fahrenheit; there were plus signs and minus signs, numbers showing universal time, Zulu time, local time, figures big and small everywhere.

Except for two evidently significant categories: wind speed and wind direction. The latter displayed no directional information at all. And the former, the wind speed, indicated zero. Plain zero. I showed Alexandra the speedometer dial, which I assumed to be broken. No, she said, grinning in a way that in retrospect I like to think was just a little sheepish. No, it was not broken at all. *There is just no wind right now,* she said. *It is a perfect calm. Zero on the scale. Smoke from a cottage chimney would rise vertically. A pond would have a mirror-flat surface, unruffled. All is still.*

She went up the iron stairway to the observation deck, where the mechanical instruments—anemometers, pitot tubes, and the like—stood quiet, static, untroubled, workless, pointless, slightly shamed at having nothing to do. She had her rubber mallet just in case there was some rime to chip off, but there was nothing but blue skies and distant clouds and the other peaks ranging southward, and likely as peaceful there as we were here at the windiest place on earth.

It happens once in a while, said her colleague, Charlie. Once or twice a year. But rather unusual. Others who work on weather statistics, he remarked drily, were trying to see if windlessness was becoming a trend.

But just then the steam whistle sounded in the distance, the signal for the cog railway's return downhill. Rather than risk being marooned on the peak, we said our farewells and shook hands and left. Nimbus looked up from his slumber with feline disdain.

On the way down I suddenly, and with a muted chortle of delight, realized the irony of it all. The underlying reason for my writing this book is the notion, still unformed and so to some a subject of considerable controversy, that wind speeds around the world are falling. That despite hurricanes and cyclones and typhoons becoming ever more violent—and indeed on the very day of this visit, September 28, 2024, a ferocious hurricane named Helene claimed scores of victims in the southern Appalachian

Mountains—despite such events, the *average* wind speeds around the world are inexplicably declining. The world is said by some to be in the grip of what they are calling the Great Stilling. Which may or may not be permanent. Which may or may not be vastly consequential for life on Planet Earth.*

And so I wondered—might what we'd just observed on the top of Mount Washington this day be an example, a harbinger of sorts? A sign? To climb to the windiest place in the world and find it enveloped in a Great Terrestrial Stilling of its own, a perfect, unusual, strange, and inexplicable calm: a more profound irony could hardly be imagined.

As the train squealed its way down to where the bare subarctic rocks sprouted dwarf pines and then full-grown tree-line trees and then deciduous species again, with yellow and orange and red and purple leaves in their glorious fall abundance, we arrived back down in the valley. The mountains rose as a great sun-dappled eastern wall above us, against a now deeper blue and darkening sky. The summit of Mount Washington was just visible, though not the buildings.

Was there a wind up there by now? Would events like today, the air silent and immobile, occur again, and more often? And if so, why? Wind, they say, is essential, invisible, and eternal. But is it *truly* eternal? Those on duty with their instruments up on the whaleback of Mount Washington will be among the very first to know.

---

* Studies published early this century suggested that between 1980 and 2010, the average surface wind speeds in continental Europe, the Americas, Asia, and Australia fell by as much as 15 percent. Out at sea, speeds remained unchanged; and within violent cyclonic windstorms—hurricanes, typhoons, and the like—velocities actually increased. What was going on? Many excitable weather scientists suggested, somewhat improbably, that the increasing "roughness" of the Earth's surface—more cities, taller buildings, regrowth of forests with larger trees—was to blame. Others put it down to the universal culprit of global warming. Meanwhile, by 2024 the situation seemed to be changing back again, with talk today of global terrestrial re-acceleration. Most remain mystified and intrigued, keeping their own counsel.

THE Breath OF THE Gods

PROLOGUE

A Feeling for the Wind

The wind bloweth where it listeth, and thou hearest the sound thereof, but canst not tell whence it cometh, and whither it goeth: so is every one that is born of the Spirit.

—King James Version of the Bible, John 3:8

Commonly shortened to: The Wind Blows Where It Pleases

{1}

However out of character that windless autumn day up on Mount Washington might have been, the very fact of windlessness is a far from unfamiliar phenomenon. There are places in the world where for days or weeks on end there are no winds at all—and most unsettling the experience of being in such places can be. Regions of unbearable calm exist particularly out on the open oceans, close by the equator, in what since the early nineteenth century have been known from the Dutch word for *dull* as *the doldrums.* A sailor marooned in the doldrums, in that notorious sink of lifelessness and airless torpor, cuts a wretched figure. Immobile, impotent, salt-grimed, half-mad with thirst and tedium, and with his craft—the "painted ship / Upon a painted ocean" that Coleridge had memorialized in his *Rime of the Ancient Mariner*—floating listless, the sails hanging like funeral shrouds, the hull pinioned to a scalding and oily-looking sea.

{2}

A robust and lustily blowing gale can be an invigorating happening, however. Ever since childhood I have listened to the nightly BBC broadcast of the *Shipping Forecast*, created a century ago for all those vessels—fishing boats, mainly—that were doing their business on great waters, as the psalm has it, out on the seas around the British Isles. Out in the ferocious gales of the North Sea, on the overcrowded routes of the Irish Sea and the English Channel, and off to the islands' west, the wide Atlantic herself. Out there, in those so-called sea areas with names that all British radio listeners would come to know intimately—Humber, Dogger, Malin, Viking, Hebrides, Trafalgar, southeast Iceland, Faroes, Finisterre—mariners would listen keenly to the near-poetic intonations of the announcers, telling them of the expected conditions—*Rockall, nine hundred and sixty filling slowly to nine hundred and ninety, westerly Storm Force Ten, backing southwest eight, snow showers becoming rain, poor becoming good*—which in this case told a fisherman who was riding the gray swells way to the west of Scotland's Outer Hebrides that a cyclonic depression was easing, the winds were slowing and backing across the compass to bring in warmer southerly air, the snow was changing to rain, and the visibility from the bridge was likely to improve in the coming hours. He would, in short, soon be a happier skipper.

And meanwhile, in common with countless others in bedrooms around the nation—for this broadcast, invariably heard around local midnight, has in recent years become required listening for a sizable fraction of the British public—I would pull the blankets more cozily up to my chin, listen to the rain drumming on the windowpanes, and reflect that at least I wasn't out in a full gale on the high seas but snug and safe at home. And it was the wind in particular that we always wanted to know about. Not the rain or the fog or the driving snow. What we always wanted was: Was it blowing Force Three or Four—the Beaufort scale we all somehow knew—or was it maybe roaring Six or Eight, with great waves and swells, or maybe even Force Ten or worse, with the tiny trawler, blinded by spume,* heeling over in constant danger of foundering and sinking to its doom? That was what made the forecast in a perverse way somehow comforting—the danger was out there, while all here in bed within, though the wind might be doing its worst outside, was secure, as it surely always would be. And so, sleep.

{ 3 }

Except—not necessarily. Strange and not entirely explicable things are nowadays happening in the world of wind. Gales seem to be blowing where they ought not to, and are ceasing to blow where for centuries past they always have. The word *unprecedented* seems lately to be more commonly used in weather reports. In the winter of 2025 the hot and dry northeasterly Santa Ana† winds did terrible, unprecedented damage, far worse than their habitual seasonal hammering of Los Angeles, when they firmly remind city dwellers of how peculiarly ephemeral life is, of "how close to the edge," as Joan Didion famously put it, they choose to live.

---

* The official description of the sea state in a Force Twelve wind—employing the admittedly highly subjective scale, by now more than two centuries old and yet accepted as gospel by mariners at sea as well as by folks in England now abed—reads: *The air is filled with foam and spray; sea is completely white with driving spray; visibility very seriously affected.*

† Like the Mediterranean's *mistral* and *sirocco*, like the Alpine *foehn* winds and the Levant's *hamsin*, the Santa Ana downslope winds seem to alter the public mood, and for the worse. Raymond Chandler wrote that when a Santa Ana was blowing, "every boozy party ends in a fight. Meek little wives feel the edge of the carving knife and study their husbands' necks." Some judges over in Provence are said to go easy on the commission of domestic violence during a mistral. As to why the strange effects: an excess of positive ions in the blowing air, some say, without any evidence.

In 2025 these infamous katabatic winds, rushing down from the desert, were said to have been directly responsible for the spread of an unprecedented cluster of wildfires in Pasadena, Altadena, and Malibu, by far the most destructive, costly, and tragic in the region's history.

Fires burning relentlessly southwestward under the powerful influence of the Santa Anas, destroying everything in their path.

The Santa Ana northeasterly winds blowing smoke from the Los Angeles fires out to sea, January 2025.

Europe had formidable outbreaks of hurricane-force winds in 2024 and 2025. And then again and somewhat earlier elsewhere, and as a specific illustration of the changing location of violent wind outbreaks, take the events of March 2023, which devastated the hitherto climatically unscathed small town of Rolling Fork, Mississippi. That was a storm of ominous aspect.

Were it not sited astride Route 61, the musically famed "Blues Alley" of the Mississippi Delta, Rolling Fork would be a place of rather limited significance. Fewer than two thousand people, 70 percent black, 30 percent white, and no Native Americans, despite this once having been Choctaw country. It is a county seat and has a city hall, a courtroom, a police station, a school, and a small depot on the Illinois Central line that once took the famous *City of New Orleans* nightly express down from Chicago to the Gulf. None of these buildings—solid structures that had stood in Rolling Fork for the past century—remain standing today. On March 24 a tornado of exceptional strength blew in from the west, leaped across the Mississippi River ten miles away and zeroed in on this sorry little town, destroying almost all of it in thirty minutes of shattering Friday-night chaos.

What made this melancholy event somewhat unusual was its location—a direct hit by a severe tornado on a town well to the east of the Mississippi River. Such was not exactly unheard of—there have been tornadoes as far east as Massachusetts—but it was sufficiently unexpected as to catch the inhabitants unawares, and also tended to confirm a growing belief among American Midwesterners that the traditional location of Tornado Alley, which runs up the prairies from Texas to Nebraska and is where most such storms occur, was shifting steadily eastward.

The scale of the disaster became starkly apparent to me five months after its occurrence, when I was at a book festival in Jackson, the Mississippi state capital. I had thought to rent a car and go see the damage for myself, since Rolling Fork was only ninety miles away. But it turned out that just about all the volunteers on duty at the festival came from the stricken town—because their homes and businesses had all been wrecked and they had no work, no homes, and nowhere else to stay. One lady, who worked for the Farm Bureau insurance company in Rolling Fork, was more than happy to drive me there and show me around.

She had been in Mexico visiting one of her sons when first she heard the news late on the Friday night. She managed to get a flight by way of Atlanta on Saturday and returned to an utterly devastated hometown that afternoon. Everything was gone. Her home was demolished, as was her office. The family's forty-year-old Airstream camper was no more than a mess of pulverized metal. The town's main water tower was down, as was every utility pole. Every government building was wrecked. Two eighteen-wheeler trucks lay upside down on top of a crushed drugstore. Seventeen people had died, and ten times that number were injured. By the time she reached town the wreckage was already crawling with police and ambulance workers—and insurance adjusters, many from her own company. She was proud to be able to say later on that she issued the first refund check on Tuesday, a scant eighty hours after the tornado struck.

Five months on, not much had changed. The Rolling Fork that I saw in August was scarcely recognizable as a place where people might once have lived—it looked like a miniature version of Gaza, only set down among the bayous and paddies of the Delta, the weather torrid and humid, its inclemency kept partly at bay by thousands of roaring generators. She and her husband were initially living in a tiny trailer that had been provided by FEMA, the government agency charged with helping after major disasters. The trailer had been brought upriver from New Orleans, where it had housed victims of the devastating Hurricane Katrina in 2005. I kept in touch with them; by now they had finally moved out of the trailer, their house sufficiently restored to be habitable. They had lived in town for forty years, but would now, shaken by the event, leave and live out their retirement somewhere more peaceable and secure.

The municipal water tower of Rolling Fork, Mississippi, brought down by an EF-4 tornado on March 24, 2023, which destroyed most of the town and killed seventeen people. Tornado Alley, it is thought, is shifting slowly eastward.

And the thing that troubled her most? The wind. "It was bad enough suffering through a tornado, which we have never once experienced before," she complained. "But ever since this storm, I feel, the wind has somehow changed. Nowadays there seems to always be something blowing up from the Mississippi River. It never used to be so windy. It rattles the windows. It makes the roof shingles clatter. The washing on the line has to be held extra secure. Things are changing, and I want no more of it. So we're thinking of leaving, heading off for somewhere the wind don't blow so hard."

To students of the phenomenon of changing wind patterns, anecdotal evidence like this proves to be both bothersome and intriguing. For however capricious the wind may be—with its caprices extending within the doldrums to long periods of seeming nonexistence, as well in higher latitudes to episodes of terrifying cyclonic insanity like in Rolling Fork—it has an omnipresence still that offers some kind of comfort. Wind, in its endless train of forms, speeds, directions, and qualities, seems always to be there, always to have been there, eternal and ceaselessly "going on," as with the beating waves of the seashore. As with life, if you will. Wind is a familiar thing, a thing whose very existence brings a kind of reassurance—yearned-for when absent, delighting when gentle, accursed when either biting cold or parching hot, feared when violent. But familiar nonetheless; an ever-present reminder of the living presence of Nature and of the planet that is so uniquely bathed in its presence.

EXCEPT, AND TO reiterate—things are changing. The reassurance offered by the simple feeling of the wind in your face is, in some parts of the world, now less of a certainty than once it was. Places in which winds once blew with unyielding inerrancy have fallen silent and airless. To be sure, and counterintuitively, the extreme wind speeds that have been measured in recent times in most of the world's hurricanes and cyclones and typhoons and tornadoes have increased, and mightily. Storms, like that in the Mississippi Delta, are getting worse, more extreme, more frequent.

But in other, more placid places, measurements of average wind speed seem to have done quite otherwise and declined—plummeted, in some cases—and there are references to areas that now experience phenomena known as "wind droughts," with environmental and human consequences ranging from displeasing to downright dangerous. A previously mentioned meteorological phenomenon—stilling—has entered the vocabulary. Global terrestrial stilling is the phrase in full, though it is sufficiently new to be missing from the latest edition of the American Meteorological Society's 850-page *Glossary of Meteorology*. In this biblically regarded volume, found between an entry for a fierce Alaskan gale known as the Stikine Wind and a series of entries beginning with the adjective Stochastic—essentially a near-synonym for *random*—there is just one brief mention of a small measuring device known as a Still-Water Level. But *stilling* is something quite different, and has only lately been brought about; and by what, no one is yet fully certain.

{4}

Wind must have been truly inexplicable to those who first felt and experienced it. Not least, of course, because it couldn't be seen. Those first sentient and sapient hominids would in time have come to understand something of the connection between sun and heat, say, or between clouds and rain, maybe even between lightning and thunder, as well, perhaps, as the notion of storms. But the unseen force that somehow bent grasses, swayed branches, tore at clothing, chilled the face, toppled structures, sometimes made it impossible to walk—how to explain and understand this invisible, mysterious, and, in its power, often frightening entity?

In those places where civilization had its beginnings, there was by all accounts plenty of moving air to go around. Almost all these early centers—regions such as the southern Indus Valley, the eastern plains of China's Yellow River, the lower Nile, cities like Persepolis, Tikal, Cusco, Teotihuacán, and of course the Mesopotamian conurbations of Babylon, Nippur, Baghdad, Mosul, and Nineveh—appear to have been awash in wind from the very recorded beginning. The inhabitants experienced then, and still experience today, robust bursts of the stuff; they bask in breezes, put up with gusts, and suffer gales, much as we do now. Early peoples in such centers eventually got around to writing about their wind-related experiences, and in doing so placed the phenomenon of wind at the very center of recorded history, since the earliest writings and proto-writings all had symbols, characters, letters, or words denoting the phenomenon, right from the very start.

And yet this very lexical fact—that wind and its kin were to be written about so early on in the story of humankind, has long been at the heart of a debate about *which came first.* Did civilized peoples notice the wind and the weather it drew along with it, and write about it? Or did the very existence of the wind and the weather somehow draw such people to those very particular places in which their civilization would grow and flourish and eventually allow them to grow in intellect, to develop systems of writing, and so to record the very phenomenon that drew them there in the first place? And just when was the idea of wind first written down, written about, somehow vaulting into its indelible position in the lexicon, central to the human story? When did wind first become a leading component of, to underline the point, *recorded* history?

Central to this early debate was a once-revered American historian and human geographer, Ellsworth Huntington, who declared that without a doubt it was weather that determined the character of those early peoples who dwelt within and among

Ellsworth Huntington, an early-twentieth-century Yale geographer, argued that peoples originating where the weather was ever changing were intellectually more able than those raised in areas of unrelenting climatic stability.

it. The cleverest and most civilized peoples, he wrote, lived in places where there was *an endlessly variable succession of different kinds of weather.* Moreover, he added, civilizations developed in places where the atmospheric environment presented constant challenges of the unexpected—and all of these unexpected events, Professor Huntington declared, were brought about, were carried and delivered, by the wind.

The book in which Huntington laid out his theory—*Civilization and Climate*—was first published by his alma mater, Yale University, in 1915, to wide contemporaneous acclaim. A revised edition came out a decade later, its basic thesis unaltered. And this, the Huntington Theory, held that when weather was varied, when seasons were identifiable, when long periods of heat, for example, were randomly interrupted by rains or cold fronts or, most of all, by wind—those who lived in such environments tended to flourish, learned how to cope with change and climatic adversity and challenge, and in consequence became equipped with a unique capacity for inventiveness, became more cunning, more open to new ideas, more cultured, more *civilized.* On the other hand, those who simply survived in scorching sand deserts or among ice floes or featureless steppes, people who were mired in meteorological monotony with little daily or monthly or seasonal variety, became, by comparison, and to put it most bluntly, dullards. Such inhabitants were certainly equipped to deal with heat or with cold maybe, but not necessarily with variety. They would rank, in Huntington's archaic and long-discounted view, as generally uncivilized peoples, unlikely to have the nous to spread their limited intellectual abilities beyond where they originated.

"Civilization seems to make great progress," Huntington wrote, "only where a stimulating climate exists. A high civilization may be carried from such places to others, but it makes a vigorous growth and is fruitful in new ideas only where the climate gives men energy." And such energy-giving climate—with moderate mean temperatures,* moderate humidity, moderate periods of precipitation of various kinds—is brought about and delivered, Huntington reminds the reader again and again, by a robust and endlessly variable passel of winds.

Ellsworth Huntington was, as might be suspected from his writings here, an unreconstructed eugenicist—no less, in fact, than a longtime board member of the American Eugenics Society. As such he is regarded today as a thoroughly discreditable fellow, his stated views on almost everything now disregarded and disdained. His low reputation today is similar to that of two near contemporaries, Britain's Sir Francis Galton—the actual Victorian godfather of the science of eugenics—and the University of Wisconsin's Frederick Jackson Turner—who spent most of his later career at Harvard—the historian who came up with the Frontier Thesis as an explanation for the unique ruggedness of American frontiersmen. All three men are much derided—which in Huntington's case would be a pity, since his work on climatic determinism has more than a little common sense about it.

And one indisputable fact remains relevant to this book: that in each of the centers where sophisticated and civilized human life did take hold—where barbarism gave way to social order and considered thought, where hunting and gathering moved aside to allow building and schooling and language and writing—in all these places there was an abundance of wildly various kinds of weather. And weather was a topic about which the local people wrote extensively. Especially, for a start, in those places where

---

* The ideal mean temperature for the development of truly civilized human behavior is, according to Huntington, around 20°C, 68°F. As it happens, and doubtless to the good professor's satisfaction, all the centers of early civilization—the Nile to the Yellow River by way of Mesopotamia—lie along the isotherm, the connecting line, that sports this particular mean temperature, with any thermal extremes all tempered by regular outbreaks of wind. Variety is the key, presenting the inhabitants with regular mind-bending challenges of the unexpected, as in a more modern device like an aircraft flight simulator.

writing itself first got underway: in the Fertile Crescent between the rivers Tigris and Euphrates in Mesopotamia.

{ 5 }

The oldest written language in the world is generally now agreed to be Sumerian, forged in the kingdom of Sumer in what is now southeastern Iraq, the origins of which date back to around 3100 BCE—five thousand years ago, during the Bronze Age. The writing is in cuneiform script, patterns of largely wedge-shaped lines that were impressed with a sharpened reed onto tablets of softened and leather-hard clay, eventually baked to ensure their preservation.

A word for wind exists in Sumerian—it is *lil*, written . The lexical story of this particular word is a little more complicated, however, since Sumerians, as far as we know, may well have been aware of wind and its effects, yet did not fully understand what caused the air—which was also invisible, of course—to move. There are Sumerian words for other features of the weather—for rain, for clouds, for ice, fog, thunder, and lightning. There is even a cuneiform word for snow, which was not common in Mesopotamia but which certainly occurred once in a while, and the existence of which, together with the challenges it presented to a settled desert people, presumably kept them intellectually on their toes, just as suggested by Ellsworth Huntington and his famous Theory. All of these things—rain, fog, clouds, snow, and so forth—are all easily seen, discernible, visible, describable, and so lend themselves to having a cuneiform noun determined for them and an utterable sound evolved for them. But with wind, only the *effects* of its occurrence are ever visible—the ripples on the river water, the waving motion of tree branches, the dust devils rising up from the sides of a desert dune, the dishevelment of clothes occasioned by a particularly violent gust. Such physical effects of wind are translatable; and so, most important of all, are the quarters from which the blasts or the breezes or the gales appear to come. The wind's direction was the most crucial lexical key, which in time prompted the inhabitants of Sumer to tell one wind from another. They gave names to these most important, most easily recognized directional winds.

The number of these winds that the Sumerians decided to name, and which unintentionally inaugurated a system of nomenclature that would be adopted by most of the rest of the world, was four. The Four Mesopotamian Winds, an invention of Sumerian mythology that has winds blowing from what are still today variants on what are known as the four cardinal points of the compass. It would be the Greek and Hebrew peoples of two thousand years later who would define the actual cardinal points, North, South, East, and West; but the Sumerians—and their Mesopotamian successors, the Akkadians, the Assyrians, and the Babylonians—worked with what they had, and that was a system arranged around their own prevailing wind directions, all of which happened to be a uniform forty-five degrees off true north. (Interestingly, a map will show that the direction the Fertile Crescent takes itself, running between its two great rivers, spears diagonally across the deserts of Iraq at the same forty-five degree angle to the north: the connection between the direction of the named winds and that of the topography across which they blow is surely more than a coincidence.)

Of the four, the most familiar, the prevailing wind of Mesopotamia, still blows today from the northwest. It blows right along the parallel valleys of the Tigris and the Euphrates, along the direction of Mesopotamia itself. In doing so, the wind helps to push along the river waters—and any sailboats that happen to be borne upon them—as they flow some 1,300 miles from the Taurus Mountains in Turkey to the salt marshes near Basra and the rolling waves of the Persian Gulf. The Sumerian word for this wind is quite simply The Regular Wind, and the deity that Sumerian mythology attaches to it is *Ninlil* (also 𒆤).

There are three such cardinal winds, each of them gods and all of them supposedly siblings, three male and the fourth, a southeasterly that tends to blow in the wet winter season, female. This wind is supposed to bring in clouds from the sea and is generally regarded as a demonic, so it doesn't greatly enhance the notion of female empowerment in what would, three thousand years later, become a predominantly Islamic society. The two other wind gods are the so-called Amorite wind that blows from the southwest and a chilly northeasterly "mountain wind" originating in the Zagros range in modern Iran.

However, the niceties of Mesopotamian wind names and their presumed powers should perhaps concern us less than one plain and unassailable fact: five thousand years ago, five words were invented from scratch. One of them, *lil*, denoted the bewildering idea of wind as an entity. Four others were born to signify the gods of the directions along which Mesopotamian air moves. The appearance of these five lexical inventions marks the first time that humankind ever attached words to this invisible and magical mystery. It was, if the metaphor may be mixed for an instant, a watershed moment. For once it had been achieved in the deserts of west Asia, so the very notion of wind as a linguistically definable entity took off everywhere. Wind words proliferated and spread with untamed promiscuity, as did wind gods, all around the planet.

{ 6 }

Egypt was probably next, then China, though there is much amiable tussling for post-cuneiform pole position. In the Nile Valley there are hieroglyphs galore that signify weather systems; and, being invisible and inexplicable, wind is denoted by its effect rather than its reality. Hence the symbol for wind in Ancient Egypt is the sail of a boat—not its hull or its crew, just a tiny mast and a tiny convexity of cloth, together illustrating the unseeable force that moves dhows along the passage between Luxor and Aswan over the Upper Nile, or Alexandria and Cairo on its more placid lower reaches.

Advance continues as word spreads. China's early appreciation and understanding of wind comes with a good deal of sophistication, a little more biology, a little more logic, something of a vaguely discernible rational approach to what still remained, at least in the Yellow River floodplain, as much a mystery as it did five thousand miles west, beside the Tigris. The Chinese spoken word *feng*, when written, has nine strokes (in the slimmed-down version used in mainland China today the character is simpler, with only five strokes). The basic etymological explanation is that the character is based on insects, since it was long thought that wind brought insects along with it; a secondary explanation holds that insects have nothing to do with the concept, but that the origin of the character has to do with birds, most notably the phoenix, and that the flapping of the birds' wings was what caused winds to blow. The explanation is somewhat charming, interesting, and thoughtful—and not simply a matter for mythology left to explanation from the heavens, as it was in Sumer. The Chinese approach has, moreover, a kind of common sense to it, even though it doesn't hold water today.

In addition, the nine- or five-stroke character for wind is, in the Chinese syllabary, a *radical*—meaning that it, like the wind itself, is regarded as of major importance within the entire structure of Chinese language and culture. It is appended to scores of other elements to provide an impressive range of other concepts, both related and little connected—as, for example, *fengcai*, meaning having an elegant demeanor, to *fenghai*, the damage caused by a storm, or *fengjing*, a ventilating shaft, to *fengmao*, a cowl-like hat worn in the wintertime.

By now—and most especially once the Romans, the Greeks,

and the Hebrew cultures had gotten hold of the idea—the notion of wind was starting fully to enter the cultural mainstream, enjoying its role at the very center of the human experience. It was no longer entirely a mystery but was seen to have uses, could perform myriad kinds of work, brought changes in weather, was by turns enjoyable and frightening, and—most significantly, maybe—worthy of deep study. And yet for many years still, some societies fell back on naming deities for the wind, keeping it as an entity under celestial control, or having unfathomable consequences—and not wishing to pry too deeply into what caused it, taking a pre-Enlightenment attitude to it, letting the mystery be.

{ 7 }

There are a very large number of deities still firmly linked to the blowing of wind, with temples and ceremonies and traditions associated in the public mind with each. Europe still has some. There is a statue in central Helsinki, for example, of the love-starved blacksmith, the eternal hammerer Ilmarinen (*Ilma* being the Finnish for air), who created very nearly everything and whose control of the wind is absolute. The Lapps have a summer-wind god, who wields not a hammer but an iron shovel. Hungary has a god whose influence has spread abroad to faraway societies where Turkic languages are spoken—among them Tuva, the strange landlocked country northwest of Mongolia where they make little more than felt and asbestos and have the ability to perform throat singing, producing two notes at the same time. There are Basque wind gods and Baltic wind gods, heavenly figures in Lithuania and Ireland, deities in Germany and Persia, the Philippines and Japan (Fujin, Shinatsuhiko, and the cartoonish Susanoo, an aeolian trinity), Albania and Egypt (the Egyptian north wind god Qebui is depicted as a man with four rams' heads). India has countless wind deities, and China has, among others, a wrinkled old crone called Feng Popo, or Madam Wind, who presides over storms and moisture and the very force that carries them across the skies.

And then, and inevitably, come the Greeks, bearing gifts. Though the Egyptians and the Babylonians had performed sterling work some thousands of years beforehand, it was the Greeks who finally got a handle on the scientific method and tradition and wrote all manner of observation and analysis of the intricacies of the natural world—wind and weather most notably. The most significant volume of scholarship in the field, still recognized today for its erudition and prescience, is Aristotle's four-volume *Meteorologica*, which he is thought to have composed in his academy in Athens around 340 BCE.

The volumes are perhaps best known for introducing the four-element concept of earth, air, fire, and water. But within Aristotle's first three books there are also abundant references to the natural movement of air—in volume 1, for example, after discussions relating to comets, dew, hoarfrost, and hail, the thirteenth chapter is devoted to "winds and the formation of rivers"; volume 2 next offers up, in three separate chapters, the great philosopher's considered views on the causes and effects of wind, the effect of heat and cold on winds, and the question of the winds' various directions. Finally, in his third volume, which is mainly given over to pretty descriptions of the formation of rainbows and haloes, he includes a section devoted entirely to hurricanes, typhoons, thunderbolts, and a phenomenon Aristotle calls "fire winds," which seem mainly to relate to the emissions from volcanoes.

This being Ancient Greece, even so logical a thinker as Aristotle defers somewhat to the influence of gods. But at the same time one can sense his attempts to wrestle his mind free from the obscurantist restrictions of the priesthood; his explanation for the origin of winds, for instance, is all to do with the differences in the solar heating of the air, the differences in pressure that this brings about, and the resulting movement of air from zones of high pressure to low—which is not too far from today's data-based understanding. Reading *Meteorologica* is to be vividly reminded of Aristotle's quite astonishing range of interests and breadth of intellect. His preference for logic and reason over dogma and faith marks him as a one-man Enlightenment.

Yet wind gods would prove quite intractable, surviving some while longer in rational post-Aristotelian Greece just as elsewhere. Probably more so. A Macedonian astronomer named Andronicus of Cyrrhus played a major role in promoting the importance of wind within the pantheon of weather phenomena, though he himself is little known or remembered today.* The most indelible legacy of Andronicus was his design and construction, around 50 BCE, of the great eight-sided, forty-foot-tall Tower of the Winds, which still stands (spruced up a little by an eighteenth-century restorer, and further cleaned in 2016) in eminently fine condition within the Roman marketplace, or *agora*, that extends beneath the cliffs on the northeast of the Acropolis in Athens. The Tower, rather resembling a monstrously large telephone booth or a public lavatory for the use of the Titans, was fashioned from pale yellow marble and decorated with flourishes of elegantly severe carvings. It sports a number of interesting architectural features (a scattering of sundials on the walls, a now-vanished water clock or *clepsydra* inside, a weathervane on its very top), all allowing the Athenian citizenry to know which way was which. And then the names—some of them now firmly annealed in various forms into the English language—of the eight gods whom the Greeks designated as being in control of the world's most significant winds.

---

* Andronicus has, however, the distinction of an admiring biographical article written by one of the great minds of the twentieth century, Hugh Chisholm, the creator of the widely revered classic eleventh edition of the *Encyclopaedia Britannica*.

Dating from at least 50 BCE, the Tower of the Winds in Athens is a robust marble structure—originally topped with a weathervane and housing a water clock—that depicts the eight wind gods in relief on its outer walls.

As a group, the Greek wind gods themselves were known as the *Anemoi*—from which we get *anemometer*, the word for the instrument employed to measure wind speed. The most prominent eight of those gods include the generally unfamiliar *Notus*, *Lips*, *Skiron*, *Kaikias*, and *Apeliotes*, the marginally better-known *Eurus*, the name for a humid southeasterly that blows up from the Aegean, and the immediately recognizable pair—the cold northerly gale named *Boreas* and the congenial westerly whose balmy Ionian breeze was colloquially known as *Zephyrus*. For a few of my childhood years our family car in England was a Ford Zephyr; we would take camping holidays in subarctic Norway, into what my father would rightly note were the boreal forests, and from clearings in which we might, with luck, see the aurora borealis, the northern lights.

One might be tempted to wonder why Andronicus settled on just eight for the number of winds and their directions, since other numbers enjoyed periods of popularity too. Homer, for example, writing eight centuries earlier, defined six winds; Aristotle twelve; and, in the Christian era, as many as thirty subtly different winds were recognized and classified. The association between wind and direction was not initially obvious; directions used by Greek sailors usually mentioned *head to this coast* or *leave that mountaintop to the right side of your boat.* It was only when navigators came to associate north winds with cold and zephyrs with (once in a while) clouds of red Saharan sand coloring the skies over Piraeus that sailors would set their sails to be filled by certain winds if they wished to head in certain directions. Then again, once the magnetic compass had been invented—by the Arabs or, more distantly and many thousands of years before, by the Chinese—life at sea became much easier, and sailors would become confident enough to sail their ships well out of sight of coasts and landmarks and rely on direction alone to speed them to places along paths that could be set by the winds in the sails and confirmed by the compass on the bridge.

By now, as the Greek and Roman Empires rose and fell and the focal points of civilized humanity shifted to western Europe, to India and East Asia, and eventually to the Americas and Oceania, wind assumed the role that it enjoys today—both as a lexical commonplace and as a concept, an essential component of cultural and economic life. Maybe not until modern times did it become a phenomenon wholly explicable and understood, but its origins and mechanism were now fast coming within reach of the capable, and in time wind was to be made great use of, turned exclusively to the forging of human advantage.

Wind soon began to appear abundantly in society's major public texts. The Bible, for instance; the Hebrew word for wind, or breath, or spirit—*ruach*—appears in the second verse of chapter 1 of Genesis: *And the world was a desolate and formless emptiness, and darkness was over the surface of the deep, and the wind of God was hovering over the surface of the waters* . . . making it the twenty-seventh word in the Good Book, coming shortly after *world* and just before *God.* After that, the word is firmly woven into the works of Alexander Cruden, the seventeenth-century Scottish divine and eccentric* whose *Concordance* to the Bible has not been out of print in Britain since 1737, found seventy-one appearances of *wind* and a further twelve of *windy*—one of them in the somewhat overwrought third chapter of the Gospel of St. John, the excessively windy epigraph at the head of this chapter.

{8}

Every English-language expository dictionary (as opposed to the much earlier bilingual dictionaries—as with Latin-to-English or French-to-English—that were the first to be so named) that has been published since the creation of the genre in the seventeenth century has devoted an impressive number of column inches to wind and its various relations. Samuel Johnson, who set the standard with his massive multivolume work in 1755, offered a definition that was characteristically *long-winded*, if I may: *Wind*, the great man declared, *is when any tract of air moves from the place it is in, to any other, with an impetus that is sensible to us . . .*

Such was the near-biblical authority of Johnson—his work was simply called *The Dictionary* by those in any household that possessed one—that his elegantly convoluted definition remained unchallenged until 1928, when the *Oxford English Dictionary*, some seventy years in the making, displaced Johnson as

* In middle age, Cruden, a notorious scold, developed an impassioned loathing for misspellings, grammatical infelicities, and wrongly placed apostrophes, believing such mistakes signified the moral decay of both the perpetrators and their country. He devoted much of his later life to erasing such horrors, traveling endlessly across London armed with a very large erasing sponge.

the ultimate arbiter of the language, for then and for all time since.

The *OED*, in its coverage of *wind*, displays how impressively the English word—which owes nothing whatsoever to the Sumerian *lil*, the Chinese *feng*, the Japanese *kaze*, or the Hebrew *ruach*—has entered the English lists by way of a wide variety of Northern European sources. After first appearing in the thirteenth century to mean, quite simply, *the movement of atmospheric air*, the word has, over the subsequent eight centuries, developed some sixty quite distinct meanings: the scent of an animal, an accumulation of gas in the stomach, a kind of fear or alarum, a type of musical instrument, the notion of empty talk, the receipt of information, the concept of riskiness, drunkenness, and so on. The principal areas in which the word finds itself employed include nautical life, astronomy, horses, boxing, music, anatomy, and the game of mah-jongg. As well, of course, as its classically spare three-word essential and literal definition with which the current *OED* begins its twelve pages of explanation and history for us: *air in motion.*

The pronunciation of the English word has changed over the years. Traditionally, the form was *waind*, rhyming with grind, rind, and mind. Only in the eighteenth century did the modern popular form develop—and that, it is believed, was to accommodate words like *windmill* and *windy*, with which the long *i* would have sounded discordant. But poets have never liked the modern polite form, not least because so few words rhyme with it, whereas there are scores—*behind*, *bind*, and *kind*, and others that fit nicely into the poetic affect—that lie nicely with the older version. Besides, linguistic purists have long grumbled about the "thinness" of the word, having no onomatopoeic connection with the sound made by a shrieking gale, whereas the romantic can surely hear the groaning of a storm that hints at the waind that is causing it.

There are also compound words in impressive abundance: the

OED lists no fewer than 269 nouns and 118 adjectives, together with a smattering of other parts of speech. Among the endless pages or their immeasurable online equivalents found in the latest edition of the *OED* one finds references to *windbag*, *windshake*, *windhole*, *windward*, *puffwind*, *wind bladder* (which keeps a fish upright), *fuckwind* (a sixteenth-century term for a kestrel, now happily obsolete), *wind-chapped*, *wind furnace*, *windsucker* (a troublesome horse afflicted with an addiction to extra-deep breathing), *wind-pinning* (the stopping-up of masonry so as to keep out drafts), *beam-wind* (a sailing term), and *winddog*, defined as the visible fragment of a rainbow, traditionally taken by farmers to suggest the coming of rain.

Inevitably there is much congenial discussion, often mildly competitive, about which language has the most words that, specifically or allusively, relate to wind. Not surprisingly, people who must endure lots of it have the richest vocabularies. The Inuit, in their various circumpolar nationalities, are already known for their having many words for snow and ice, but are less well known for their lexical fondness for wind. In the Arab world, with its desert dunes and sandstorms, wind is the basis of an extraordinarily rich vocabulary. Much the same is true in Sanskrit and Japanese. But the most promiscuous language of all in this regard is undoubtedly Hawaiian, which is said to have well over *six hundred* recorded words that can fairly be described as wind-related. The remote island chain in the mid-Pacific is bathed in what were until recently endlessly regular northeasterly trade winds—but so wildly complex is the topography across which these winds blow that all manner of local variations are experienced, and have for centuries past been noted and catalogued by a Hawaiian people with a legendarily profound interest in, knowledge of, and affection for their environment. For every single *ahupua'a*, or geographically distinct area across the islands, there are three quite different winds—more even than the hundreds of names for rain, and for which Hawaii is already justly famed.

Wind, in short, seems a universal. Air in motion finds its way into just about every activity and inactivity of man, beast, plant, and thing that exists in the world above its waters—and since, as we shall see later, the connection between wind and waves is intimate and of immense importance, it works its way and its will on the surface of the waters too. The wind brings and it takes; it slows things and by turn it speeds them up; it interrupts and it hurries along. It lifts seeds and supports birds and insects. It warms and it chills. It builds and creates; it ruins and destroys. It can be trivial or it can be catastrophic, interfering with plans for picnics and scattering fleets in battle. It directs explorers and settlers to lands to be colonized or dominated. It shapes human geography. Air moving across reeds, single or double, produces music of ethereal beauty; sounded through brass it resonates and stirs to action. It generates power, flies kites, performs work, lifts water, and drives watercraft—thousands of different shapes and numbers of hulls with innumerable sail designs, from jennies to spinnakers, that seek the best ways to speed from place to place, powered without cost by the invisible and eternal motion of that cocktail of oxygen and nitrogen gases that allow the world's land creatures to survive. Wind alters the moods and attitudes of human beings—in France one can still beg a court's mercy for the malevolent *mistral* that impelled you to commit a murder. In Los Angeles a winter wind can lead you to imagine the darkest of thoughts. The word appears in the titles of a thousand books and quite as many films, whether it blows in the willows, suggests a vanished way of Dixie life, is against, is mighty, or is inherited. In China the wind is called the Sigh of the Sky. In Greece there is a peaceful morning wind called the *aura*. More than four hundred winds around the world have names, though the previously noted six hundred specifically cataloged in the Hawaiian vernacular suggests there may be very many more than that.

But only wind's consequences are visible, not the wind itself: you see the turning sail of a mill, the perilous lean of a close-hauled yacht, the storm-broken limb of an ancient oak, the flattened heads of a field of barley, the truck blown onto its side where the highway crosses a valley. For those who in the past were content to leave wind to its own devices, to its gods and to the flexible explanations of mythology, a measure of inexplicable invisibility was just fine. But now, the Enlightenment has blown away the cobwebs of dogma, here as in cloisters and apses around the world; science has stepped in, has come to grips with the highly complex physical realities of wind and weather in all its myriad forms—meaning that at last we can now come up with a fair set of answers to these nagging questions: Just what is the wind, and why on earth does it blow?

{9}

Of all the ways in which the various components of the world's weather systems present themselves, the manner in which wind does so is surely the most compelling, the most widely experienced, and—at least in lexical terms—sports the longest history. Rain and clouds, hail and snow, blistering sunshine, rolling thunder, falling frogs, dense fogs, sleet and ice storms can each be memorable and spectacular—but they are all, whether extreme or inconsequential, only fleeting phenomena. It's true that in a few corners of the world particular weather elements are so dominant* as to render all others insignificant. In other instances (the Indian southwest monsoon springs to

* In many parts of eastern India, in the states of Assam and Meghalaya, rain is decidedly *not* occasional, being well-nigh constant, night and day, with up to 460 inches falling each year. In the small hill town of Mawsynram, supposedly the world's rainiest place, the inhabitants incorporate enormous straw umbrellas into their clothing, giving them the appearance of two-legged tortoises. The air that brings all the rain originates in the Bay of Bengal and is brought to the Himalayan foothills by ceaseless southwest monsoon winds.

mind), the onsets are sufficiently regular and relied upon that they have become central to timekeeping, to the understanding of seasons and of natural cycles. But wind more generally has the property of near constancy, rather than being merely occasional, just an *event.*

Moreover, and despite being technically just one of the many moving parts of all weather systems, wind has somehow contrived to gain an equivalence to the sum of all the other elements combined. Think of the long-employed phrase *wind and weather*—an alliterative warning, first written of more than a thousand years ago, of a sustained onslaught of atmospheric misery. In Charles Dickens's *Dombey and Son* there is reference, for instance, to a sturdy London mansion that had been built "proof against wind and weather." Dickens would never have written of proof against *fog and weather* or *snow and weather.* Wind alone, at least in the English language, has long assumed the same heft as the sum of all the other elements that comprise the weather.

Wind has broad shoulders, it has power, it has influence like no other. And it is sustained, is always there—faint, maybe, or just around the corner or over the horizon, or else strong enough to bring the clouds and hurry the rain and, in time, to corral the snow into enormous drifts. It brings, it takes, it drives along all the other elements of weather; it underpins the main components of the world's climate. Which is surely why humankind for so many thousands of years past has noted the existence of this initially quite inexplicable phenomenon of *the movement of the air.*

It will never be known for certain whether early human societies—Sumerians, say, or Chinese, or those in the valley of the Indus, or the Nile, or the Ancient Greeks—devoted their first concentrated studies of the phenomenon to one particular kind and strength of wind in preference to another. Did it seem more prudent and sensible to note, to classify, to assign words and phrases to those movements of air that were initially thought of as *manageable,* as reasonable, as moderate? Or did some especially prescient early village philosophers or riverbank-dwellers suppose they might one day harness some of these natural air movements to perform work, and so devoted their time and energy to classifying winds that one day might be thought of as useful, as potential working winds? Or again, might some braver souls decide to consider those winds that were downright dangerous and alarming and difficult to work with? Aristotle certainly did, never shying away from any hurricane that might lash his corner of the Mediterranean. Or else did early students of the natural sciences decide on the easiest option of all—on those breezes that seemed most open to explanation and understanding, and which in a practical, observable sense did little more than stir flowers and rustle leaves and create waving fields of corn, so easy on the eye?

Or did they maybe consider all this together and look at the movement of air as ranging across a spectrum, shifting the atmosphere in ways from the calm to the catastrophic? Quite probably they did exactly that—took the phenomenon of wind as a whole, as a thing of many kinds and variations, of differing velocities and forces and powers and directions and temperatures and effects. And if that was the case, as I suspect it was, then these early students would surely have begun their studies by examining winds at the most benign and congenial end of the spectrum. Maybe by addressing winds that were pleasing to deal with, in time it might be possible to work out why they blow in the first place.

CHAPTER ONE

Light Airs, Gentle Breezes

When the oak is felled the whole forest echoes with its fall
But a hundred acorns are sown in silence by an unnoticed breeze.

—Thomas Carlyle, *Signs of the Times*, London, 1829

{1}

It is a crisp, clear, and quite cool early spring morning in 2016 on Wyoming state highway 14A, some sixteen miles northeast of the small city of Cody, and there is a dusting of late-season snow still on the distinctively shaped knoll of Heart Mountain. The peak, which from a distance does have the vague outline of a human heart, rises abruptly out of the rolling plains of the Bighorn Basin. It is a site of much geological interest, since it once stood a hundred miles south, close to what is now Yellowstone Park, but got here near to Cody fifty million years ago in what some say was the world's largest landslide.

There is a small museum here telling a more melancholy local story of the concentration camp that was hastily thrown up nearby in 1942 to house thousands of Japanese Americans, detained on the orders of a panicky government for no good reason soon after the attack on Pearl Harbor. They were held without trial here and in nine other similarly remote sites around the country; one hundred twenty thousand people in total nationwide, some fourteen thousand of them in Wyoming, briefly making the Heart Mountain Relocation Center, as it was officially known, one of the largest cities in an otherwise thinly populated state.

Of the sprawling camp, little remains. After the war was over in 1945 and the hapless inmates were sent away with twenty-five dollars and a Trailways bus ticket home (and no hint of a government apology; that would come grudgingly forty years later), most of the barrack huts were torn down and sold to local farmers for a few cents on the dollar. Nowadays the land is otherwise covered with *Artemisia*, blue basin sagebrush, which puts out a nice clean-kitchen smell after a rain shower. One surviving structure, the commissary (or, according to some, the hospital), sports a brick chimney, maybe sixty feet tall and rather rickety, not likely to stand for many more years. My wife and I were there taking photographs, and we chose our spot for the picture we would take the next day such that the chimney rose out of the sage in the foreground and the immediately recognizable stump of Heart Mountain stood five miles away at the back. On this crystal-clear morning the image we planned seemed little short of perfect.

But it was not to be. To our dismay, the following morning Heart Mountain was quite invisible, as if while we were sleeping a thick gray blanket had been dropped in front of it. And you could smell why: it was smoke from burning trees. Probably from pine trees, since there was a vague oily note to the otherwise acrid woodsmoke. It became clear in seconds that Heart Mountain had been enveloped that very morning in a particular kind of smoke: thick Canadian smoke that, as the local Cody radio station soon reported, had been brought stealthily down from the wildfires that for the previous week had been chewing away at the boreal forests up in central Alberta, a thousand miles away to the north.

It so happened that during the night the wind had changed,

both in its direction and its strength. The day before it hadn't been blowing perceptibly at all. Now, at ground level, there was just the faintest and lightest apprehension of a breeze—Beaufort One at best, I thought—a light air that barely stirred the desert brush. But the direction was key. And during the day all kinds of observers, human and electronic, had been plotting the direction, the course taken by what was, after all, a swathe of inbound windborne pollution.

A study performed the year before when smoke from burning northern Canadian forests had first been noticed as far away as Baltimore, causing widespread irritation, both actual and metaphorical, had employed an immense array of impressively named technologies—Tropospheric Ozone Lidars, ceilometers, radio spectrometers, a spaceborne device known as CALIOP, or Cloud-Aerosol Lidar with Orthogonal Polarization—and hosted more devices of increasingly arcane nature. The overall purpose of directing this battlefield-scale range of hardware toward something so modest as a trail of smoke was—because smoke can be seen, whereas unpolluted air generally remains invisible—to understand more fully the physics of what is now called the planetary boundary layer, the lower couple of miles in which the atmosphere most closely hugs the Earth's surface, and where most people live and most weather happens. The PBL is home to all our microscale weather, and a good deal of its senior relation the mesoscale too; but synoptic systems seldom pass this way, and it is into the lesser categories that wildfire smoke and the winds that bring it are placed by weather professionals.

In the 2015 study, the smoke plume was more complicated than it would be a year later. It originated in the same northern forests, was picked up by the then-blowing northwesterlies and swept down across the uninterrupted prairie farmlands of central Saskatchewan and southern Manitoba, passed over Winnipeg, crossed the Great Lakes and entered the United States near Chicago, turned slightly to a more easterly direction, and stalled in the Ohio River Valley in the middle of a high-pressure doldrum for a day or so before clambering adroitly up and over the Appalachians and heading into the maw of a battery of instrument-wielding agencies and universities—NASA, NOAA, Johns Hopkins, Goddard, the University of Maryland—who, unbeknownst to the wind, had been studying it ever since it left home a few days before—after which further inspection and the smoke it carried with it headed out across the ocean and vanished over the edge of the American world.

The 2016 wind, *our* wind, had carried its smoke in a more direct manner, was stronger and faster, and performed its work with much less fuss. It had first come to public attention—indeed, to world attention—because it ripped through and well-nigh destroyed Fort McMurray, a northern Alberta city of eighty thousand relative newcomers that was dominated by the oil and gas industry. (The Athabasca Oil Sands, long known to contain immense quantities of hydrocarbons but which have only lately been made economically extractable, are nearby. The fires did not touch the oil; had they done so, it would have been a wholly different story.)

The wind at the time of the fire and the city's consequent near-total evacuation blew from the north. The cocktail of gases and smoke particles that forest fires help create was thus immediately picked up and sent on its way in a more or less due south direction. Wherever it passed it covered in a blanket of gray invisibility. It did so as it crossed into Saskatchewan, then as it swept over the US border into Montana, and after some three hundred miles of steamrollering across Big Sky country between the foothills of the Rockies and the start of the High Plains it arrived in Wyoming and the sagebrush scrublands around Cody and Heart Mountain. Unimportantly, it quite ruined our day. The fact that the appearance of this windborne miasma of woodsmoke spoiled the taking of a picture of an American concentration camp was trivial compared to its other effects—car wrecks on highways, exacerbation of children's allergies, complications with ozone, darkness at noon, and so on.

{ 2 }

It also served as a reminder that even the gentlest of winds possesses not only speed and direction but pressure and force. It has power, and with that power it can pick things up and carry them to far-off places, with all manner of unanticipated consequences. In the northern summer of 1883 the volcanic island of Krakatoa, in what is now Indonesia, blew itself to pieces, with the wind then distributing the trillions of tons of dust that had been thrown up into the sky all around the planet. The effects were truly astonishing: sunsets became exceptionally vivid—in Poughkeepsie, New York, firefighters were sent galloping through the streets to put out what they were told was a huge conflagration in the northern suburbs but which turned out to be a flaring sunset colored by windborne Javan volcano dust. Edvard Munch painted *The Scream* with the Norwegian skies a swirling mess of purples and oranges and bile greens, thought also to be Munch's memory of the dreadful skies that were seen in Oslo in late 1883, when he first decided to paint as he did.

Dust carried only in the troposphere—like the smoke from Alberta that fetched up in Wyoming—will settle back on the Earth's surface in fairly short order, especially when the wind transporting it dies away. But powerful volcanic eruptions can hurl the very lightest morsels of dust right through the troposphere and up into the lower reaches of the stratosphere, where they have been known to remain suspended for long periods of time—up to ten years in some cases. And once the particles reach the stratosphere, the jet stream can get to work and set them swirling around the globe for many years after the event that first put them there.

Whether briefly suspended in the troposphere or endlessly held farther up, windblown dust can have dire effects on the weather below. Benjamin Franklin, a most prescient man, noted a strange dry fog that screened the sun and chilled the air in the summer of 1783, somewhat discommoding him. He put it down to the dust from an eruption of the Icelandic volcano of Hekla, "and that other volcano which arose out of the sea near that island, which smoke might be spread by various winds, over the northern part of the world." Winds blew similarly heroic quantities of dust from the 1815 eruption of Tambora, shortening wheat-growing seasons in Kansas and Russia, causing immense and prolonged rainstorms across Europe, and making life so miserable for Mary Shelley in Geneva that, it is said, she was inspired to write the gloomiest of all her stories, *Frankenstein.*

And more recently, in late April 2010, the sustained eruption of the hitherto-unknown Icelandic volcano Eyjafjallajökull filled the European airspace, once again courtesy of the jet stream, with so many needle-sharp spikes of ash that almost one hundred thousand flights had to be canceled and millions of travelers around the world found their lives suddenly upended. I was at a dinner in New York for the heads of two dozen University of Oxford colleges and watched as, one by one, their cell phones pinged to tell them they were now stranded in Manhattan indefinitely because plumes of erupted volcanic dust had suddenly put their aircraft engines at risk.*

* Most famously a British Airways Boeing 747-200 with three hundred passengers and crew flew through a high-altitude ash cloud eight miles above Java in June 1982, with all four of its Rolls-Royce engines failing in midair. After a difficult half hour of involuntary downward gliding, the cockpit crew managed to restart the engines and land safely in Jakarta. The engines were clogged with melted volcanic debris, and the windshield and landing light covers had been sandblasted into opacity. When a relieved flight engineer kissed the airport tarmac and responded to his captain's asking why by explaining that "the Pope does it," the captain returned, "Well, he flies Alitalia."

{ 3 }

Much that is nearly invisible also gets blown and distributed by gentle winds—germs and bacteria and bacilli and pandemic-causing viruses, all manner of unseen unpleasantness can and does hitchhike on the balms of what we generally like to think of as *fresh air.* In doing so, however, most germs' efficacy becomes diluted, and, at least over longish distances, they are rendered largely ineffective as dangerous contaminants. The same cannot be said, however, of one deeply insidious breeze-borne passenger—radioactive fallout—and a new word describing a numberless community of affected victims was coined in 1982: downwinder. *A person who lives or has lived downwind of a nuclear test site or reactor, where the risk of being affected by radiation is greatest.*

Those who worked on the Manhattan Project at Los Alamos during the early 1940s well knew the theoretical dangers of radiation and the likelihood that an atomic explosion would result in lethal fallout—that debris blown upward by a bomb and contaminated in the process by radiation would inevitably be pushed by the winds far from the site of the detonation itself, probably causing casualties much in excess of those directly affected by the blast. It goes without saying that there was a great deal of fallout produced by the first two bombs used in anger, which wholly destroyed the Japanese cities of Hiroshima and Nagasaki. But subsequently, as first the Soviet Union and then Britain, China, France, India, Pakistan, Israel, and a host of other countries acquired nuclear weapons, so an orgy of testing got underway—and with the prevailing winds at all the test sites proving to be unwitting conspirators, so millions of people found themselves, either at the time or, more commonly, very much later when symptoms started to appear, victims of the terrible maladies caused by radiation. And in nearly all cases the downwind victims were the poor and powerless, whose early complaints went unheeded.

Those affected by windborne radiation in the United States had more of a voice than most; and the downwinders of the far western states, close to the Nevada Test Site, won—and are still winning—compensation under the terms of the Radiation Exposure Compensation Act. But those in Kazakhstan who suffered grievously from the carelessness of the Soviet nuclear operators of the Semipalatinsk Test Site; the Australian aboriginal peoples hurt by fallout from British testing at Maralinga and Emu Field; the Polynesian islanders unlucky enough to live under the irradiated clouds that swept over them all too frequently from the French test atolls of Mururoa and Fangataufa; the Uyghurs who lived near the Lop Nor desert test site in China's far western Xinjiang province; and those peaceable villagers who had spent their years fishing and making coir mats on Bikini and Eniwetok atolls in the Marshall Islands—all had cause to blame, until they later came to know otherwise, the winds that brought them such terrible illnesses and caused the gruesome deformations of so many of their unborn children.

All Japan still knows the case of the *Lucky Dragon 5*, the 140-ton fishing boat that, with its twenty-three crew members, was contentedly fishing for shark and tuna some one hundred miles east of Bikini Atoll on March 1, 1954. Those of the crew who were up on deck at 6:45 that morning saw the sudden fierce fire-glow of what we now know to have been the infamous Castle Bravo test of the world's first hydrogen bomb, a test that was wildly larger than the US government had anticipated. The fishermen, though totally awestruck by the sight, supposing it to be a solar phenomenon (even though the sun itself was busily rising on the other side of their horizon) were not frightened, and were initially unaffected by any blast wave from the bomb. Whatever had occurred might have been spectacular, but so far as they could see, it wasn't dangerous.

But then, some six hours later, something began to fall from the sky. Coming down from the cloudless heavens—and doing so in the middle of a calm and waveless blue sea—what we now know to have been the fragments of roasted coral and melted sand drifted down in extraordinary abundance, like a blizzard from nowhere. All of this material had evidently been brought to where the boat lay drifting aimlessly while the nets were doing their work below by the prevailing and sweetly scented westerly wind.

Not unreasonably, the crewmen were interested by this unusual phenomenon and collected some of the fallen debris with their hands, in one case tasted it, and had no thought that it might be dangerous. But it was, and terribly so—and by nightfall all the men were sick. They were developing blisters, their facial skin was turning dark, their gums were bleeding. They promptly pulled up their nets and began to head for home. As their voyage northward progressed, so they became more and more unwell, with dizzy spells, stomach problems, and appalling headaches, and in time most started losing clumps of their hair.

Suspecting the proximate origin of their problems, the men stuffed a cotton bag full of the fallen snow-ash and took it with them in their sleeping quarters, a decision that a later inquiry confirmed made their situation even more dire. Panicking, they hurried back to Japan, making as good progress as they could to their home port of Yaizu in southern Honshu—their arrival delayed for several days by a storm in the western Pacific—and reached the quayside in an ever more deteriorating condition. When they reported what had happened, they were greeted with astonishment by a wholly unprepared local hospital.

In time the twenty-three were moved up to Tokyo, where—because of the detonation of the atomic bombs nine years earlier—there was some idea among doctors of how to treat patients with what was now realized to be acute radiation poisoning. One of the crew members, the chief radio officer, died soon after his arrival in Tokyo, suffering horribly; the others recovered only very slowly, remaining in the hospital for well over a year. And throughout the process the Japanese doctors tried in vain to win some practical cooperation and advice from the American Atomic Energy Commission experts—but almost none was forthcoming. American doctors did indeed arrive in Tokyo to observe the treatment offered to the men, but it seemed to all the Japanese staff at the hospital that the Americans were there to study the victims rather than to help them. And there was criticism too that the senior Japanese government officials involved in the affair deferred to their American colleagues. One crew member, Oishi Matashichi, complained in a subsequent book that Japan's foreign minister was acting "as if he was America's foreign minister rather than our own."

This was a singular marine tragedy. It had been brought about by an American bomb test that went badly wrong. But it was a tragedy visited upon the hapless Japanese ship by a prevailing (and, as it happened, incorrectly forecast) westerly wind. The historical importance of the incident was the realization by Japanese and British radiation scientists that the men had suffered from poisoning not by the products of a "conventional" atomic bomb test but of a top-secret test of a more lethal and powerful thermonuclear device.

And so, against the advice of what history has since shown was a cravenly pro-American Japanese foreign ministry, the doctor who realized this, in concert with British scientist Józef Rotblat,* made an angrily formal announcement to the effect that for

* Rotblat, originally from Poland, joined the Manhattan Project in 1944 in the belief that a bomb was needed to counter Nazi Germany's nuclear ambitions. But once Germany had been defeated, he resigned from the project on grounds of conscience and said he felt betrayed by the bomb's use against Japan. He spent the rest of his life campaigning for peace and against atomic weapons, and is regarded as one of the main architects of the Partial Nuclear Test Ban Treaty of 1963. He was a firm believer in World Government.

the fishermen to have been affected by these hitherto unfamiliar radiation products, America must have tested a brand-new type of bomb. The cat was out of the bag—and the ultra-classified test had been revealed because a high-altitude wind had spread the far-larger-than-expected radiation plume a far greater distance than anyone had imagined. Such is the unanticipated and metaphorical cleansing power of some kinds of wind—in this case revealing a closely held American military secret and telling it to the world.

By 2025, it was widely believed that all of the crew of the *Lucky Dragon 5* had succumbed to a variety of radiation-linked ailments. The author Oishi Matashichi, who had spent most of his later life writing and campaigning against nuclear weapons died in 2021. He was eighty-seven. With the passing, in 2024, of one further elderly officer from the ship it was assumed that all of the crew had now passed away. But then in the spring of 2025, more than seventy years after the incident, it was learned that an eighty-nine-year-old civil servant in the ship's home port of Yaizu was still alive. Like most of his fellow crew members he had suffered much social ostracism from the belief of many that radiation sickness was some kind of contagion; he had tried to live out his later years in quiet obscurity.

His boat, now thoroughly cleansed of all radiation products, is on public display at a Tokyo museum. The United States government begrudgingly paid more than $15 million in compensation. The westerly winds in that corner of the Pacific Ocean still blow on, invisible as always.*

* In response to US president Joe Biden's approval in 2024 of a new and beefed-up American nuclear strategy directed primarily at China, one critic noted that the jet stream would certainly bring any plume of radiation from a US attack on the Chinese mainland right back to California, with dire consequences. An "own goal," in soccer terminology.

{4}

And then, eight thousand miles and a whole hemisphere away from the Pacific's equatorial westerlies, there are the southeasterly winds that blow regularly over the Swedish mainland in the springtime. On the morning of Monday, April 28, 1986, one such breeze, bringing with it a light rain, was wafting over the village of Forsmark, in southern Sweden, two or so train hours away from Stockholm. Workers were lining up to begin their week at the community's only large-scale employer, a three-reactor nuclear power station that at the time suppled an eighth of Sweden's electricity.

It was routine for workers at the main gate to check themselves for radiation as they entered the plant. As one employee, an early arrival, casually placed his boot under a scanner that was pointed down at the neatly clipped lawn, a banshee wail suddenly sounded, alarming everyone. It was a radiation detection alarm. The grass outside the plant, it turned out, was highly radioactive. As were worker after worker who then found themselves also contaminated. The plant's operations manager promptly shut down the reactors and performed an elaborate safety check to ensure there was no leak from within the plant itself. There was none; and by mid-morning the reactors were burbling away as usual, without incident.

But additional reports from elsewhere in Sweden cascaded in, all indicating there was radioactivity all around, in a swathe running from the southeast to the northwest of the country. Clearly the wind was spreading whatever was in the air. Within a matter of a few hours, it was determined by Sweden's Radiation Safety Authority that the air's deadly cargo had not originated in Sweden itself. Since the wind was coming from the southeast—and with Forsmark located on the shore of the Baltic Sea—it seemed highly likely that the cloud of whatever radioactive material was triggering the alarms had to have originated on the far side of the water.

Two Baltic states, Latvia and Lithuania, lay on that far side, and a straight line would pass across and through the Iron Curtain and into the Soviet Union proper, through what is now Belarus and into what is now Ukraine. The event that brought about this frightening discharge of radiation must, the Swedish government concluded privately, have occurred in the southwestern part of the USSR, fully one thousand miles away, somewhere near the Ukrainian capital city of Kyiv. Moreover, as the day wore on, Swedish radiation scientists were able to analyze the components of the radiation, concluding that yes, with radioactive isotopes of iodine and cesium having all the recognizable characteristics, the radio signature was undeniably that of an old-fashioned Soviet-built reactor. There had surely been some mishap, and somewhere in the southwest of the sprawling USSR.

But nothing had been said publicly—even as ever more radiation was detected spewing onto Swedish territory and, now that the word was somewhat out, falling on detectors in Norway, Finland, Denmark, the Netherlands, West Germany, and the northern extremities of the British Isles. Stockholm made formal contact with Moscow. What on earth, their diplomats asked, is going on?

The response from the Kremlin was unambiguous and clear and to be expected. Nothing untoward had taken place. The Soviet Union was in all senses running just as it should. Nothing to see here. Please move on. Kindly don't bother us again.

But the wind doesn't lie, and with geography and meteorology and radio-isotope distribution creating a trinity of unchallengeable facts, the Swedish government made the rare and brave decision to go public. A spokesman informed the press of what had been observed at Forsmark and then across the country, and told of the officially suspected reason why. I recall the moment well: I was making a cup of afternoon tea in the kitchen at home in Oxford and switched on the 3:00 p.m. BBC news. "Reports from Sweden," the newsreader intoned with unusual solemnity, "suggest that some kind of accident may have occurred in the Soviet Union, most probably at a nuclear power station in Ukraine. The Soviet authorities have said nothing."

The effect was, in Soviet terms, swift, if not exactly electric. Late that Monday afternoon Radio Moscow broadcast a brief item, a tacit admission, saying that there had been an incident early on Sunday morning, and then using as the event's location the name that would soon become infamous around the world: *Chernobyl.* Soviet missions in both Sweden and West Germany

In April 1986, unseasonal southeasterly winds blew telltale radiation from the Chernobyl nuclear site toward detectors in Sweden. Had they been blowing from the west, as is usual in springtime, the USSR might have been able to deny the event had occurred.

then made formal requests, if unannounced and expressed *sotto voce*, for advice or help in extinguishing a reactor fire. Whatever had taken place was evidently gravely serious. There had been casualties, Moscow admitted.

The wind in central Europe then backed eastward, and radiation detectors in Kent and Hampshire in southern England began to chatter with unacceptable levels of pollution, and later that Tuesday the displeasing cloud spread farther south onto the Channel Islands. Europe started to get restive. Children were advised not to drink milk. Warnings were broadcast: vegetables needed to be washed, fruit peeled. People should leave the eastern parts of Soviet Union as soon as possible; embassies and travel agents went into overdrive. And still the fires at Chernobyl raged on, and ever more dangerous radiation was vomited out of the ruined Reactor 4. Radio Moscow used the term "disaster" in one broadcast, only to have an editor quickly erase the word. The official response of the Soviet government seemed confused, panicked, ineffective. Evacuations were ordered, exclusion zones set up. Millions, then billions of rubles were soon being spent on mitigating the unprecedented environmental damage caused by the accident. The saga wound on for weeks, then months, then years—and Chernobyl still commands attention around the globe today, for being one of only two Level Seven nuclear accidents in world history (the other being the generally well-managed Fukushima disaster in northern Japan in 2011).

The wider historic implications of Chernobyl were hugely significant. At the time of the event, the Soviet Union—led by Mikhail Gorbachev, the first figure to have been born after the state's formation in 1917—was undergoing rapid political changes. There was heady talk of more openness and economic freedom; and yet here, with this lethal disaster at a Ukrainian nuclear power station, the government lied and prevaricated and obfuscated just as it had done for years in the past—and the Russian public, it swiftly became apparent, was not going to tolerate such behavior anymore. A cascade of events followed—and were maybe triggered to some extent by—what had happened here. The solidarity of the Warsaw Pact, the physical integrity of the Iron Curtain, started to fray. Hungary openly criticized Moscow's handling of an accident that spread deadly radiation across its border. Poland, already in turmoil over labor reforms, was likewise publicly unhappy. Three years later the Berlin Wall fell, and after a series of defections of Pact countries from Moscow's mothership—Lithuania first, Kazakhstan last of the fifteen that had put the *U* in USSR—the party was over. The Soviet Union was consigned to the history books, a wholesale political and economic failure. What had begun with the Bolshevik Revolution of 1917 was finished in 1991—and as Mikhail Gorbachev was later to say, incautiously in the views of many, it was Chernobyl that triggered it all. The accident "made me and my colleagues rethink a great many things," he said later.

And the public awareness of Chernobyl was first brought about by the prevailing southeasterly winds, which made all of Scandinavia—well beyond the strictures of the Iron Curtain—sit up and take immediate notice. Had the winds that spring weekend been westerlies instead, the radiation plume would have spread entirely over Soviet territory, and Moscow could have kept a lid on the story for many months and maybe years after. But winds, knowing no political boundaries, swept the story into the public square, with consequences more profound than anyone at the time could have imagined.

{ 5 }

Fierce winds can blow down trees and sink ships and cause immense amounts of physical damage. But gentle winds can cause damage too, since all winds possess the ability to lift and carry a variety of items into the sky. However, in saying such a thing we find ourselves on the lip of an arithmetical crevasse, for the detailed mechanics of the wind's potential power can be daunting, and a common reader—just like a common writer—can sink ever deeper and more intractably into a morass of higher mathematics, with differential and integral calculus most probably present in all its confusing glory. A standard 1976 American undergraduate textbook—*The Ceaseless Wind* by John A. Dutton—summarized this challenging reality by noting that such processes as the carrying power of winds of various strengths are all part of *the complex chain of processes in an atmosphere that is forced into ceaseless motion by thermal effects on a rotating planet.* It is prudent to recall such summaries whenever a chasm of incomprehensibility looms and you peer over its edge.

At the heart of the matter, however, is a single comprehensible physical fact, which simply holds that *the pressure exerted by wind is proportional to the square of the wind's speed.* Given this relationship between speed and pressure, three further points of definition need to be noted: the wind's *force*, its *power*, and the *work* it is able to perform. This pressure (a quantity measured in pascals) is directly related to force (a quantity measured in newtons), and both of these qualities determine, in a mathematically definable manner, how much work (a quantity measured in watts) a wind of a certain speed can perform. Whether, for instance, using as an example the Albertan smoke-carrying winds of 2015 and 2016, a wind can pick up, countering the accelerating force of gravity, a tiny particle of what en masse constitutes smoke, keep it and its colleagues aloft, and push the entire group along from where they were collected to where they are eventually to be set down. Mathematics will demonstrate the complex set of principles behind such a transport; whether such explication belongs here is a matter for argument, but I have decided it doesn't. Francis Beaufort, I suspect, would have made much the same choice.

Two centuries ago, Beaufort devised his very particular scale mainly for his fellow mariners, describing how intensifying airstreams affect the elemental maritime entities of sailcloth and seawater. In more recent times, biologists and botanists, human geographers and geologists have followed suit, noting in detail how gathering winds affect the subjects of their individual fascinations. Smoke figures in one column—Human Activity—of what is now referred to as a Biological Wind Scale, invented by a famously idiosyncratic South African writer-scientist named Lyall Watson. Each upward notch of his scale is relatable to the students of a variety of disciplines. Since this section of the book is devoted just to *gentle* winds, I won't cross the line—the *biological wind threshold* is its informal name—beyond which winds are more destructive than playful. The effects of these more ill-favored winds will appear in chapter 5.

As it happens, smoke appears on this scale's gentle end, and as an entirely benign entity. Lyall Watson probably could not imagine a plume of darkly polluting smoke extending from a raging forest fire a thousand miles through rural Canada. Instead, he speaks of a single column of blue smoke rising from a cottage chimney. In the windless environment of his Force Zero, the column rises vertically, quite undisturbed, conjuring up an image of an elderly couple inside their cozy little home, taking tea beside their first fire of the year, which they had set with kindling and crumpled newspaper because the late afternoon weather outside had for the first time, and so early in the autumn, turned a little chill. A scattering of red and yellow autumn leaves had already drifted down from the maples and settled onto the mirror-smooth surface of the pond at the end of their lawn. Fall was coming, though with a graceful sense of calm.

Come the arrival of Force One, however, and there is a slight but symbolic change. The smoke now "drifts," writes Dr. Watson. Nothing more. If the couple inside gather their cardigans around them a little more snugly, aware of a faint draft seeping in under the door, we are not told, and can only suppose. And anyway, that is the last we shall hear of smoke, for the next entry and all subsequent entries in the Human Activities column relate only to how we as sentient beings respond to the rising wind around us. It is at Force Two, for instance, that we first feel the wind on our face. Then, at Three, "dust is raised"—this being the point when the "gentle breeze," its pressure now related to the square of its speed in feet per second, creates a force strong enough to lift a dust particle from the ground and pass it into the lower level of its atmospheric surroundings. Perhaps at Force Four we notice that our hair is disturbed and that looser items we are wearing may flap a little. It is only at Force Five that we experience directly the consequences of the Force Three dust—because at this point, according to Watson's scale, we risk suffering "eye discomfort from airborne matter." In other words, the dust raised by the gentle breeze is carried high enough and held there for long enough by the fresh breeze that is Force Five that we get it in our eyes and so begin to curse a wind that no longer seems quite as mild as it did some forces back. But here we are approaching the red line, when with the human effect of a Force Six "Strong Breeze," according to the table's author, our "arms are blown out from sides." Really? This seems improbably dramatic, and is perhaps no more than a fancy, an example of Watson's occasional mischievous, hyperbolic fantasies.*

---

* Malcolm Lyall-Watson (he dropped his first name and hyphen on arrival in London from his native South Africa) had a puckish wit. He was an eccentric and truly polymathic biologist, anthropologist, zoo owner, and ethologist, telegenic and eloquent, who explored outlandish places, made strangely watchable documentary films, wrote two dozen books, and then in later life (he died in 2008, aged sixty-nine) did his reputation few favors by befriending the Israeli psychic spoon-bender Uri Geller, bringing a stable of sumo wrestlers to perform in the Albert Hall, arguing that plants could hear the agonized cries of cooking shrimp, lecturing about a Venetian child who could turn uncut tennis balls inside out, and maintaining a ready ear for paradoxes, most famously that *if the brain were so simple we could understand it, we would be so simple we couldn't.* His thought-provoking creation of a Biological Wind Scale, mostly reasonable but for the flailing arms, demonstrates his appetite for the interesting and the unusual.

Watson next applies his scale to whole plants. Not surprisingly, they remain quite unmoved during periods of calm and are sturdy enough to resist any pressure from light airs, but at Force Two some leaves begin to rustle, then twigs start to move, then small branches, after which—when the breeze has whipped itself up to a fresh breeze—small leafy trees begin to sway entirely, until finally, before the threshold of destructiveness, large branches start to move and creak ominously.

{ 6 }

Incorporated into the swaying of his "small leafy trees" there is, however, one oddity, an outlier: the appearance in the scale of the behavior of that quintessentially American specimen of botanical mystery, *tumbleweed*. The rolling of a monster tumbleweed—which can only take place when the breeze rises to the level of *fresh*—allows us to look in a little more detail at the process of the dispersal of seeds in general by virtue of the wind. It has a name: anemochory.

Aside from triffids, plants tend not to move, other than, in most cases, vertically as they grow. They have limited mobility. A few species do manage to creep along the ground laterally and spread themselves and their offspring by such means, but most rely on external forces to disperse the seeds by which they are propagated from the parent plant's otherwise firmly rooted home. The agencies that help spread the seeds around are either living creatures—birds and plant-eating mammals—or nonsentient but moving entities, either water or wind. All the dispersal mechanisms have a common aim: to get the seeds as far away from the parent plant as possible, since to have the plant simply drop its seeds and have them unmoved next to itself will result in the new growth jostling for the essentials of space and sunlight, a contest in which no one wins, no one thrives. Distance is key, and the wind is an excellent tool for providing it, effortlessly. Oaks may drop acorns vertically in the fond hope that animals will snap them up and scurry away with them for winter snacking; witch hazel may squeeze its seeds out so forcefully they can briefly fly through the air at almost 30 mph; *Hura crepitans*, a tall tropical tree known as the sandbox tree, produces fruit that distributes its seeds by exploding with such a bang that it is colloquially known as the dynamite tree; and the more common phenomenon of what is called *explosive dehiscence* gives a robust shove to some seeds, like those of the charmingly named squirting cucumber. But the efficiency of these methods pales into insignificance when compared to the sheer efficiency of wind dispersion. Whether the seeds are engineered to float in the wind or designed to flutter slowly to the ground, it works, and the passage of seeds so transmitted through the skies tends to entrance all who see it in action.

With the very lightest seeds, the air can fill with wafting clouds of them, almost akin to the pollen clouds that erupt from chestnut trees in the early spring, microscopic male gametes on their way to fertilize the female part of the trees to make the very seeds that then lend themselves to the vectors that will take them places. Pollen is powder, like smoke, like haze, the components seldom visible to the naked eye. Seeds can be seen individually, can be tracked, caught, weighed.

Our gardener's seven-year-old daughter, Hazel—as suitable a name for a gardener's child as can be imagined—gladly accepted my commission to collect the floating seeds of a poplar tree: five dollars for a hundred of them. In due course the child solemnly handed me the Ziploc bag with its enclosed mass of the fluff she had snatched from the sky. With tweezers I carefully removed a single seed with its tiny attached parachute—its *pappus*—and attempted to weigh it. There was no response from the most sensitive scale I could find. I piled on nine more. Nothing. Only when I had heaped on the entire contents of the bag did the readout register: two grams. Divided by one hundred, it was possible to suggest that each seed weighed in at only 0.02 grams—and so even if the air was filled with them, an entire eventual football-field-full would weigh no more than a cup of flour. Small wonder poplar trees are so common and so widely spread; it is the policy of a poplar to manufacture as many seeds as possible and cram the air with them to better ensure their dispersal far and wide, and allow the wind to do as it will to further the process.

Dandelions behave in much the same way—a near-spherical collection of featherlike pappi, their suspended seeds clustered at the top of their stalk, waiting for the wind to blow strongly enough to separate them into their component parts and then allow each to drift, a tiny parachutist, to the point where the breeze drops below its ability to offer support, where it drops to what its parent hopes will be germination-friendly ground. Human children—and some adults in lovestruck moments—offer a small amount of help by plucking the entire gossamer sphere from the ground and blowing the seeds away. The more adept, performing for the gullible, will calibrate their number of individual puffs of air to mimic the time of day—performing the trick at dusk, say, taking six puffs to move all the seeds into flight, the coincidence suggesting that the dandelion somehow *knows* the time of day, the only flower to be a natural timekeeper. Few fall for it.

As it happens, though, there is some other evolution-related interest to the dandelion, suggesting that some of its subspecies do "know" more than might be supposed. On some small islands the local dandelions are themselves all miniature, the length of their pappi much shorter than normal, designed in a way that is much more suited to the short flights they will encounter, since they have almost nowhere to go. Island winds blow mostly over the engirdling sea, where not even the hardiest dandelion will grow.

Drifting and wafting is not the only manner by which anemochory is successfully achieved. Aerodynamics plays a part in some species, with evolution cleverly creating tiny airfoils that help the seeds to fly more efficiently with the wind, though with the same intention of getting as far as possible from the location of their parent tree. The maple, common enough in our corner of New England, creates a winglike structure bonded to the seed in what is called a samara.

When a maple samara, damp with rain or still wet with youth, is dislodged from its branch, it will drop straight to the ground. But when in the early autumn it has dried out and its single delicate wing is quite translucent, the wondrously curved surfaces create a true airfoil, said by arborists to be, in airflow terms, technically more efficient than even an aircraft wing or a helicopter blade. It is detached from its parent's crown, and as it drops it now whirls around, its concavities allowing it to autorotate, creating a miniature spinning tornado above itself that lifts the falling seed upward, allowing it to remain in and with the wind for lengthy periods, and so be carried along opportunistically to wherever the breeze happens to be passing. In the case of our maple trees, the samara's seed is at one end of its wing; in other trees—sycamores, willows, and elms, for example—the seed is in the middle of the wing, giving the whole assembly the appearance of an old-fashioned clock-winding key. Indeed, English children spend many an hour collecting fallen sycamore keys and hurling them up into the air and watching awestruck as they helicopter their way with the wind, landing scores of yards from where they were thrown. Spinning jennies, they are called in the north country—and their curious aerodynamics enjoy as much attention among flight enthusiasts as do boomerangs, frisbees, and skipping stones.

And then there came, as Watson had noted, the movement of tumbleweed.

{7}

In late 1879 the steel rails of the Chicago, Milwaukee & St. Paul Railway reached the tiny prairie town of Scotland, in what is now Bon Homme County, South Dakota. The original intention of the railroad's backers was to win business from the exploitation of gold that had newly been found in the Black Hills of what was then western Dakota Territory. Small eastern territory towns like Scotland, which sported nothing but plains and grassland and slow-moving rivers, with not a troy ounce of precious metals to be found anywhere, were happy to lay on a watering stop for the passing steam trains—on which, occasionally, came freight shipments for their inhabitants, almost as an afterthought.

One day—the precise date is now lost to history—a local hardscrabble pioneering farmer growing sorghum and alfalfa on the open prairie just outside Scotland decided he might make a better living by switching to a newly modish crop that he had heard fetched good sums on the Chicago exchange: the multipurpose grass known in Linnean terms as *Phormium tenax*, otherwise generally known as flax. By some accounts the farmer's name was Henry Schatz, though considering what happened next it seems a little unfair to single out one figure who may well still have relations living out in those western parts. Whatever his name, this man ordered a single bag of seed from a grain merchant in Minneapolis; he was told, since flaxseed at the time came from exporters in tsarist Russia, that it might take some while to reach Dakota Territory. No problem, responded Mr. Schatz, who paid his few dollars for a hundred-pound bag, thereby confirming the order.

After some months, the bag duly arrived. It had come by sea, of course, most probably from Odessa to New Orleans, thence on a barge up the Mississippi River to St. Louis, thence by a variety of railroad trains up to Sioux City, Iowa, crossing the then territorial frontier to the big junction at Yankton, and finally northwestward across the grasslands to Scotland, to be dumped unceremoniously on the tiny cottonwood platform. Mr. Schatz, if indeed that was his name, hoisted the heavy bag onto his buggy and took it up to his already plowed few acres of pasture. He took the better part of a day to broadcast the unfamiliar flaxseed into the furrows, and then let the rain and sunshine work their magic as he made a plan for reaping the profits from his brand-new crop.

Instead, botanical mayhem ensued. We will never know if the Schatz flax fields took off, or if he and his heirs and successors prospered as they had hoped. What we do know is that secreted among the flaxseeds were a small number of soft, round, yellowish pellets, the seeds of a plant hitherto quite unknown beyond the dry steppes of central southern Russia, and known ever since as the prickly Russian thistle, *Salsola tragus.*

This is a plant that dives right into the soil of any land that has been disturbed—newly plowed fields, construction sites, areas around new highways or railways—and nothing could be more

A single tumbleweed sphere can contain a quarter of a million seeds, which will be spread as it is blown across the landscape.

suitable than an area of prairie where the native grasses have been scythed away and the ground torn up by metal plowshares and the hooves of a farmer's horses. And so, when the smattering of the small yellowish pellets of *S. tragus* found themselves lying amid the crumbling black loam of the Great Plains, they seeded themselves, sprouted, and went happily mad—creating what more than one professional botanist has called "the fastest plant invasion of America's history."

Such an invasion might not be so bad if the plants concerned were benign, nutritious, sweet-smelling, pretty, or all of the above. But the prickly Russian thistle is a perfectly monstrous piece of botanical grotesquerie, loathed by just about all. And it is now everywhere, for the simple reason that for the last 150 years it has been relentlessly distributed, and with supreme efficiency, by the one commodity with which, along with open space and big sky, America's Great Plains are most liberally supplied: wind. Huge, endless, powerful, omnipresent wafts of warm and silk-smooth curtains of all-carrying wind.

For prickly Russian thistle becomes tumbleweed. That is to say, tumbling weeds themselves are not uncommon, and tumbling appears as a dispersal mechanism in not a few species of plants. But none is so large, so obvious, so emblematically Western, and, in a perverse way considering how it is so universally loathed, so *romantic* as the tumbleweed of this prickly Russian import.

At first, settling into the moist black loam and looking around itself, it seems an innocent-enough sproutling, a soft and gentle little plant, harmless, probably edible, quite attractive to browsing Herefords. But very swiftly it changes its looks and its character: after just a few weeks it becomes woody and stiff, thorny, unpalatable and tough, and, rapidly, enormous. Nothing now liked to eat it;* farmers cursed it as it dug its thorns into

---

* Although during the Dust Bowl years of the mid-1930s—covered more fully in chapter 5's pages about strong and destructive winds—some near-ruined farming

the legs of both horses and cowboys. It soon became impossible to plow, as lines of the great goblet-shaped bushes, each three feet across, formed impenetrable natural fences in the paths of the tractors. As the year goes the plant puffs itself up into a roughly spherical shape, until, come the first frosts of late October, the whole plant snaps off from its roots and waits for the wind to start it rolling across the landscape. Now it has at last transformed itself into tumbleweed proper, ready to sidle and bounce and roll its way wherever the wind chooses. Lyall Watson has his Biological Table start moving tumbleweeds at Force Five, a strong breeze; people who live in Utah and Nevada and New Mexico today insist that the tumbling starts at a much lower threshold, and that the whole rural world seems now to be rolling in unison all the time, a wretched nightmare of a body-snatching invasion. And since each sphere—though itself quite dead—holds within itself up to a quarter of a million still-fertile seeds, it leaves behind it a train of potential new prickly Russian thistle plants, which will spring to life no matter what the condition of the soil on which they land.

Hollywood used to have a decided fondness for tumbleweed imagery during the heyday of the Western movie. Few were the opening title scenes that evoked the loneliness of the cattleman or the hard masculine world of the Plains—with its creaking windpump, the skull canted by the gulch, the distant butte, a landscape pricked with sagebrush or saguaro cactus—that went without a tumbleweed or two rolling sedately across the screen.

That romance has long since evaporated. In today's world, with the American West no longer lonely and unpopulated, the wind-scattered tumbleweeds have become a menace—or at least

---

families found that as prickly Russian thistle alone grew in their quite devastated and soilless land, they might as well try to cook and eat it. People today speak of surviving the very worst years on a diet of "canned tumbleweed."

have collided, quite literally, with what some might consider the equal menace of modern housing developments. Where such gatherings of mansions and exurban refuges for those eager to escape the winter chill of New England or Canada have risen, the winds still blow uninterrupted and bring with them pepperings of sand and dust—and, nowadays, tumbleweeds, in vast numbers. The cul-de-sac—a favorite of those who plan these housing developments—has created artificial catchments for the bouncing balls of highly flammable weeds; and many are the mornings in recent years when residents awaken to find their houses and their cars and their exits covered and blocked by ten-foot-high mounds of spheres of dead and vicious thorns. The West can seem to a shivering Bostonian an alluring place; but its wildness, with its rattlesnakes, its scorpions, and now—ever since Bon Homme County, Dakota Territory, 1879—its tumbleweed, can winnow away the allure all too quickly. And in the case of the spreading of the noxious and ill-loved prickly Russian thistle, it is all nowadays the fault of the ceaseless prairie wind.

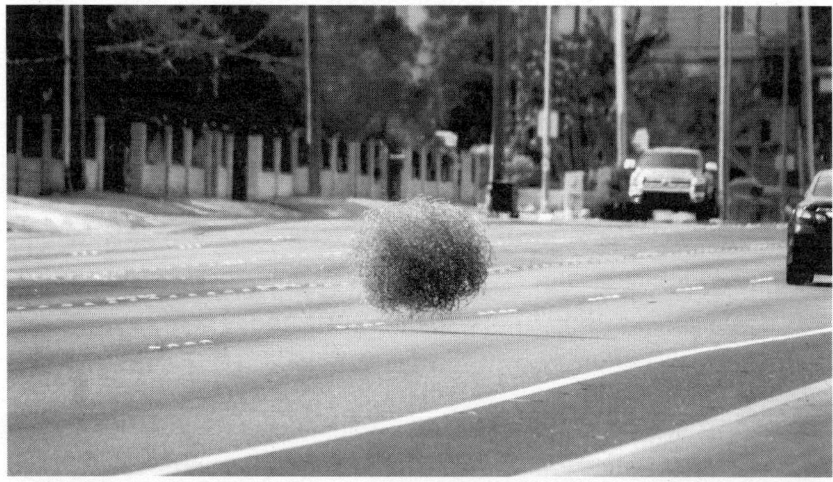

A century after tumbleweed's accidental introduction onto the Dakota prairies, wind now blows squadrons of the bouncing beasts deep into modern American urban developments, blowing them across highways and piling them up roof-high, trapping homeowners.

{ 8 }

Moving air makes sounds. Or, rather, we think it does. What in fact makes the sometimes harmonious, invariably evocative patterns of sound when a wind is blowing is the air moving past and so rubbing itself against immobile things—as with "the yellow fog *that rubs its back upon the window-panes*"—and thereby creating, one hopes and on occasion realizes, some kind of breeze-borne euphony. Windowpanes might be found by Prufrock in T. S. Eliot's imagined cities, true; but out in the landscape, the wind does indeed move itself alongside a vast array of static things—against great and barely moving and creaking trees, booming up beside vertical cliffs and through whistling sea caves, sculpting sand dunes and creating a susurrus as it drifts energetically through miles of wild grasses.

Tim Dee was for thirty years a BBC radio producer and poet, an avid bird-watcher and walker, a thinker who supposed among a multitude of other things that the different types of walking had some influence on the kind of lines a poet might write—Coleridge a hurried scrambler (his wholly over-the-top "Eolian Harp" most appropriate to cite here, and illustrative of Dee's theory), Wordsworth a measured lingerer, whose "I Wandered Lonely as a Cloud" (sighting wind-affected daffodils "fluttering and dancing in the breeze") is similarly apposite, and similarly demonstrates the same idea.

Tim Dee had long been fascinated by the wind, and the perceived sound of it. They have a term in the recording business—wild track, the ambient enveloping sound that all engineers are obliged to record during radio interviews—most especially when recording outdoors and wind becomes an inevitable component of the ambient sound. It was customary for anyone recording to ask the guest, after the interview was over, to please be quiet for thirty seconds while he held his microphone high in the air simply to record background sounds—the wild track, the sounds where the wind might happen to be.

The idea of wind as background, or the wind as essentially soundless unless it rubbed past something fixed, intrigued Tim Dee. He knew that air passing through a winter pine tree makes something of an angry sound, while air gusting through a leafless birch tree is much like a late-night jazz drummer working his brush on a snare. Telephone wires sing, barbed wire growls. But what if there was nothing for the wind to rub against? What if the choir invisible was inaudible too, did not exist; what if the wind was singing to itself, and solo?

So he sought and won a modest commission from a small London documentary film company, a firm with sufficient reputation to well-nigh guarantee a slot on BBC Four, a free television channel that specializes in the unusual and intelligent. And with that, Dee set out on a brisk November morning with his trusty fluffy-dog-on-a-stick microphone—more exactly a Rycote Windjammer, near-universally used by radio producers working outside—and his Swiss-made Nagra recorder to see if he could capture the sound of the day's strong northeasterly breeze, pure and simple, in a place without anything that the moving air might rub up against. He chose that strange three-sided indentation of the Lincolnshire coast in East Anglia called the Wash. In walking some seaward miles out there he left behind the world of trees and hills and cottages and trekked into a wholly uninhabited world of endless horizons, of total flatness, a place without woodlands, without meadows, without hills—just low marsh grasses, tongues of mudbanks with rills and swales, and by way then of a slow tidal inching into the shallow and waveless edges of the North Sea. Once in a while there were interruptions, of course—American bombers roaring high above, using the mudflats of the Thames estuary for target practice; and sliding along a tidal channel a duck hunter in his elderly boat, his fowling piece at the ready, the pleasant stink of his pipe briefly sweet on the gusts. Generally, though, all was quiet—except for the birds, those creatures who, almost uniquely, have the ability to master the wind.* So there are geese and shelduck, pipits and curlews, plovers and skylarks, all soaring or nesting or migrating or settling as the multitude of weather systems for which the Wash is known collide and multiply in the winds around and above them all.

There is a low hillock at the very edge of the land, looking like a Saxon burial mound but more probably made by modern man, a project to bend the waters to his will, Canute-style. And it was up at the barrow's flat top that Tim Dee sought and found his wind, as pure and elemental and uninterrupted as it is possible to find. Up there, no more than sixty feet above the meadows and the marshes, there was nothing but the wind. No Air Force, no rolling thunder, no show, no awe. No brent geese, no squawking chatter. No sea sound, no waves that otherwise we might have heard munching and chewing away at the edges of England. Dee raised his Rycote Windjammer to the sky, pointed it to the northeast—*Force Five*, the BBC had forecast that dawn, reminding us of old Beaufort once more—and kept himself silent to capture the blissfully pure sound on the Nagra. It was the wind talking into its own ear, the sound a bird might hear as it wafted on its lonesome, high up in the sky.

It was a congregation of many things, containing airs in multitudes of moods—a soft, thunderous hum, a rippling tuning note of white noise, a low whistling like a distant waterfall. It

---

* Tim Dee relates a haunting childhood memory of a tragedy that involves the birds' natural mastery of the wind. His newspaper delivery bicycle route took him over the Clifton Suspension Bridge, which crosses the Avon Gorge, a bird-filled wind funnel. It was a late November afternoon, and the young Tim saw a man in front of him catch his eye and then suddenly, and with near-balletic grace, vault over the safety barrier. For just a second the wind seemed to catch him and carry him back and upward before releasing him to fall to his death. Meanwhile, the birds all around kept flying, rising and falling, unaware. It made the youngster realize that the air here was a place that belonged to the birds; it "was not ours."

seemed to me to hold at once notes and pitches and tones, rolled into one seamless entity, that elevated it from mere noise, or simple sound, into something quite ethereal. Tim Dee noted that in all likelihood his microphone inside the furry dog—and yes, we knew, as did he, that the wind was rubbing against its membranes and polymer backplates and diaphragms, and that the sound we heard was not as pure as crystal, only as pure as was practicably attainable—was the first thing the wind had encountered in its thousand-mile journey from the cold front that first seeded it somewhere off Norway.

And he also wondered out loud, after immersing himself for many minutes in what then seemed to him a near sacred sound, that perhaps the dead might be wafted away, eventually, into the wind. Which is a thought shared by many ancient cultures, not the least being the Polynesians in the South Pacific Ocean, of whom more later.

{9}

Commerce tries to replicate such things. Wind chimes, of course, have been around for centuries. Aeolian harps too, configured in a thousand shapes, nowadays grace costly lawns in many a suggestible suburb, no doubt with attendant problems from neighbors hoping for peace and quiet. Nor do soothing wind sounds come cheap: one model that has titanium strings will set you back $1,654, a rather over-specific sum that must surely guarantee you the peace that derives, according to the instrument's accompanying booklet, from "the infinity of the cosmos, perfectly adapted to induce a state of deep meditation and inner silence."

A far more intimate connection with the cosmos might be made by listening to the wind not working its way through wires of precious metal but between the often equally musical grains of ordinary beach sand. Though perhaps not exactly ordinary;

those sands that have musical qualities tend to have a larger grain size than is customary—as large as half a millimeter in diameter. But if certain conditions are met—if the sand is dry, if it is made of silica (coral sand doesn't work), and if it moves under the influence of a gentle wind or a firm footfall, then it will emit booming, whistling, thundering sounds—it will *sing*, as romantics insist—as it shifts itself constantly into the dune formations that are generally associated with the world's sand deserts.

The notion of musical silica fragments has long made its way into popular culture. Bing Crosby had a decent hit in 1942 with "The Singing Sands of Alamosa," set in Great Sand Dunes National Park in southern Colorado.* And a decade later the reclusive Inverness spinster Elizabeth MacKintosh, writing under the name Josephine Tey, created one of the great unremembered detective novels of the century, *The Singing Sands*, with an epigraph that turns out to be a vital clue to the cause of the initially unexplained death of a man found in a cabin of an overnight Scottish sleeper express: *The beasts that talk, / The streams that stand, / The stones that walk, / The singing sand . . .*

Sands sing to public acclaim in more than three dozen well-known and windy places around the world. They sing—or squeak, or howl, or boom, according to your taste or your memory—in the two topographical environments where sand most commonly accumulates: dunes and on beaches. Colorado is one of the more musical expanses of dunes; Barking Sands in Hawaii would be another, except the US Navy uses it as a missile test range; local tourist bodies like to advertise the choral charms of the great sand dunes (and famous sacred-document-filled caves) in Dunhuang

---

* The distance from Cincinnati, where I happened to be, to Great Sand Dunes National Park is 1,280 miles, taking the average motorist a little more than nineteen hours nonstop. There being few places of interest in the plains of western Kansas or eastern Colorado to detain the passing stranger, it turned out to be a journey, one very long and full day, that I still remember all too keenly.

in western China, and along the Skeleton Coast in Namibia, southwestern Africa, sand deserts in both of which I have traveled quite widely but have heard not a peep. Beach sands are quite another matter, and many seem to whistle and creak and groan much as it promises on the packet, and do so particularly in places such as the Philippines (Palawan), the Inner Hebrides (the Isle of Eigg), and Singing Beach outside Manchester-on-the-Sea, Massachusetts.

As to precisely why the sands here make noises, in fine detail it remains something of a mystery, other than being explicable in the most basic and maybe rather obvious way by noting that, under the influence of wind—and, to a lesser extent, feet—*sands move.* The grains jostle against one another, surfaces scraping and shearing against surfaces, and in much the same way as a dampened finger passed around the lip of a well-made wineglass can set off a self-amplifying musical note, so moving sands can set up a harmonic pitch with an often surprising purity. The pitch of a wineglass note, like that of Palawan pure white beach sand blown by a constant easterly breeze, has been measured at around 450 Hz, corresponding to the B above middle C, an auditory pleasure wave. Some students of sand suggest that, rather than the sound being generated by the scraping of grains against one another, it results from the compression and decompression of the air trapped among the particles. And such sound-making does not invariably produce pleasure; it might sometimes cause pain. Some of the more thunderous desert sands, especially when there is a sudden slumping of a dune face caused by an unexpected gust of wind, are said to trigger booming sounds at volumes of up to 105 decibels, which, considering the dangers to humans of exposure for sustained periods to volumes greater than a mere 80 decibels, seems either mercifully rare or wholly improbable.

"Sand is a sort of snow that never melts," wrote the much-honored English meteorologist Sir Napier Shaw in his famously idiosyncratic 1933 book *The Drama of Weather.* His implication held that under the endless influence of wind, grains of sand, just like flakes of snow, manage to infiltrate everything, pile up in extreme masses, block passage, destroy buildings, bring down trees, and prevent farmers from farming, livestock from eating, or humans, unless properly equipped, from getting around.

The seasons, however, bring a welcome physical change of state to snowfalls. The regular onset of warmer springtime weather causes the snow to melt, the water's wintertime solid phase to be returned to its customary liquid phase, and then to be absorbed as liquids can be, either into the unfrozen ground or up into the gaseous atmosphere. With such changes duly made, the many challenges of winter snow start to vanish, and come soon to be regarded as only a matter of temporary inconvenience. Sand, however, does not enjoy the benefit of any phase change, no matter the environment that envelops it. Sand is a solid, if granular, and yet in its behavior is much like a liquid. It trickles through your fingers. It piles up in waves and ripples. It can be made to mimic in appearance the surface of a sea. In fine-grained form it can drizzle through a narrow opening between two bulbs of glass at a predictable and unalterable rate and so be used to record the passing of time—as in an egg timer or a sand clock. But what sand never does is either evaporate or be absorbed by the earth on which it accumulates, nor does it ever disappear. It moves, just as snow does, under the influence of the wind; and in so doing it has, over time, an immense influence on geography, on human and animal existence, on patterns of life.

{ 10 }

The figure perhaps most closely associated with the study of how wind configures this ineradicable silicate substance is a well-born Englishman and soldier-hero, Ralph Bagnold. He is to sand as Francis Beaufort is to sea.

In strictly military terms, Bagnold had great fun in the desert. Perhaps in his entire career. He first joined the British Army as a sapper, performing the mundane tasks of building bridges and docks and making roads, proving so adept that his superiors packed him off to Caius College Cambridge to study engineering. They then dispatched him to Egypt armed with a now formidable array of nonlethal skills—how to navigate with a sun compass, how to fix any kind of gasoline-powered engine, how to read and write in Morse code, how to defuse a land mine. Though he was occasionally detached for short soldiering spells in India, the Egyptian desert was the place he found most suited to his temperament, skills, and fascinations, and by the mid-1920s he was on a 3,700-mile sponsored expedition in the Eastern Sahara, fully immersed in the world of sand and leading explorations in a small fleet of Model T Fords that had been specially modified to function without undue trouble in the roadless, waterless wilderness of the North African desert.

Brigadier Ralph Bagnold, a British army engineer, devoted his life to studying desert sands and the formation of windblown dunes. A crater on Mars is named in his memory.

His early achievements as a desert explorer were legion and resulted in a bestselling book, *Libyan Sands: Travels in a Dead World*; his knowledge of the behavior of windblown sand and the creation of dunes placed him at the forefront of a small gathering of geographers and geomorphologists entranced by such arcana. His profound knowledge of the topic would come a little later and as a result of illness—a classic illustration of the adage about ill winds and the dispersal of fortune. On one of his journeys he was felled by a stomach ailment then known simply as *tropical sprue*, was returned to London, and temporarily invalided out of the service. But during his recovery he was encouraged to set up camp at a well-funded London college where he was able to build a wind tunnel to observe at close quarters what the grains of silicates did under the influence of breezes and gales and the like, and steadily came to understand the arithmetic of it all. Once fully well he was then sent off to perform fieldwork in what was his idea of the Elysian Fields—the empty vastness of the Libyan desert. He took off to what is still named the Great Sand Sea, what is otherwise unheroically known as an *erg*, and there he mapped and surveyed on a grand scale those shapes and slopes and angles and measured the shape-shifting movements of wind-driven dunes in real time, all to confirm what he had predicted and extrapolated back in the Imperial College wind tunnel.

The importance of Ralph Bagnold's work was to be born in London in the late 1930s, between the laboratory benches of Kensington and the Libyan oases of Gilf Kebir and Jebel Uweinat. But before he could publish the work for which he would become acclaimed around the scientific world—*The Physics of Blown Sand and Desert Dunes*—war broke out once more and he was recalled to the army and sent, not unnaturally, back to the Sahara and to the most romanticized and heroic chapter of his ninety-four-year life.

With the enthusiastic approval and support of the then head

Ralph Bagnold founded the Long Range Desert Group in 1941, operating behind German lines in Libya. His truck-borne troops spied on and raided enemy installations while amassing much data on sand and wind.

of Middle East Command, General Archibald Wavell,* he set up what was to be called the Long Range Desert Group, with a remit to work behind German and Italian lines and spend its time harrying and marauding and gathering intelligence and generally spreading dismay to the enemy all around. This was where the "fun" came in—a mad little group of quasi-military madmen blowing up fuel dumps, clearing minefields, sending

* Wavell, who went on to become the penultimate viceroy of India before Lord Mountbatten presided over Britain's withdrawal and the creation of Pakistan, put together a decidedly nonmilitary anthology of his favorite poetry, *Other Men's Flowers*—260 poems meant "for men to read in the trenches," as he put it, which has seldom been out of print since its first publication in 1944.

false messages to ensure Italian forces went on wild-goose chases, stealing ammunition from secret desert stashes. When Major Bagnold, as he then was, set up the group, he decided that he wanted it to be composed of a small corps of men who were strong, imaginative, self-reliant, mentally as tough as nails, and able to operate in wild and severe landscapes wholly unsupported and possibly for very long periods. New Zealand farmers, he concluded, would be the most likely to match these criteria, and accordingly the group was generously seeded with Kiwis, and the main memorial to their wartime service is now in a suburb of Auckland. The exploits of the LRDG, as it is still remembered, form the basis of an all-too-typical postwar film, *Sea of Sand*, with what was a then comfortingly familiar cast, including Richard Attenborough and John Gregson.

Sea of Sand, though an eminently forgettable movie, does serve as a reminder of the nature of the book that won Bagnold—by now deservedly promoted to colonel—his stellar reputation. For as Sir Napier Shaw had suggested years before, sand is a solid that behaves like a liquid—flowing, rippling, building into waves, even drowning those who become carelessly mired in its more fluid forms. And just like the sea, it is vastly influenced by the wind that bears down upon it, and invariably in mathematically predictable ways.

The shape of what the wind creates—sand dunes, basically—depends to a great degree on the size of the sand particles it encounters. Constant wind has a way of sorting out the sizes of the tiny bits and pieces it comes into contact with, much as a farmer uses the wind for winnowing his newly threshed grain. Where I live in New England one still sees large old barns with great open doors facing westward, toward the prevailing wind, and a similar pair at the other end to allow a continuous stream of air through the building. Within, the threshed grain, corralled on the barn floor by the thresholds, has already been separated from the straw, and now the wind is employed to winnow the corn from the chaff, the grain from its husks. Whereas the threshing is a violent act, often involving flails or horses' hooves or automobile tires passing over the bundles of newly harvested crops, the subsequent winnowing is a more gentle affair, the crops being tossed up into the air with the wind blowing away the chaff and at the same time separating—a little crudely perhaps, but nonetheless quite usefully—the better and heavier grain from the lighter and dustier fragments, which the wind will transport a few yards farther away. Close by the feet of the winnower will be a rising pyramid of corn or wheat or barley, its shape determined, as Ralph Bagnold would calculate, according to the grains' most singular property, their various *angles of repose.*

For as with winnowed grain, so with blown sand. Wind separates, files, classifies, sorts—blowing the lighter material more distantly, retaining the heavier components closer to their point of origin. It is much the same with water: a classic example of aqueous sorting is to be found on Chesil Beach in Dorset, an

Dunes, barchans, and sand ripples form a classic windblown Saharan landscape.

eighteen-mile barrier beach of pebbles that vary in size with an extraordinary degree of precision—those at the southern parts of the beach being quite hefty, up to four inches across, while the pebbles at the northern end are tiny, smaller than peas. The gradation in size along the eighteen miles is steady and impeccable, such that a sailor or fisherman landing on the beach at night can tell immediately where he has landed by the size of the stones—daylight seldom proving him wrong.

The size-sorting of sand grains occasioned by a steady desert wind, or by an onshore coastal breeze, is key to the building and shaping of dunes. Consider: A stiff breeze is unleashed over a sand-filled plain, setting the tiny spheres and spikelets of silica settled on the surface jostling and then rising and moving in the wind's direction. The lighter grains, the sandy equivalent of the dustier fragments on the farmer's threshing floor, will be blown fastest, rising and dropping down and rising again; they will bounce, or, to use the word geomorphologists have lately borrowed from the Latin, they will *saltate*, leaping or dancing along the plain.

The coarser and heavier grains, however, will behave differently. Rather than dancing merrily along, they will lag behind, able only to roll or creep under the impress of the breeze, doing so until that critical moment when the wind and the grain it propels encounters some kind of obstacle in its path—a small, immobile stone, say, or a patch of vegetation. The wind weakens a little or becomes slightly turbulent, and the propulsive force lessens, stopping the sand grain dead in its tracks so that it falls to the ground. In doing so it will then become an obstacle itself, halting the wind and those heavier particles that are coming up behind—the lighter grains advancing from the rear like eager infantrymen will leap or bounce or saltate around the obstacle, but the obstacle itself will become steadily larger and larger, will slowly become a dune. And usually, in these circumstances, a crescent-shaped dune—an accumulation of sand that is fat and tall in the middle and has long tapering wings that stretch ahead on either side. The whole structure, which can rise to many feet high and many yards long, has come to be called, from the local name in the Turkestan where they were first catalogued, a barchan.

The vocabulary of dunes is extensive and bewildering; in some cases the words are derived from the Arabic, since those hardy souls who travel across deserts know the different types and geometries of the immense structures up and over which their camels patiently plod. So, there are seifs and draas joining the barchans, there are pebble-specked plains called serirs, ready for the barchans' advances, and in the Sahara the northeasterlies that shape them and blow them there are the harmattans. Other terms have nothing to do with Arabia: the stoss side of a dune is that smooth slow-sloping side up which the sand rolls to the summit, the slip face is the steeper lee side, where the wind is sheltered by the dune's own mass, and the slope of the face is determined by the angle of repose of the kind of sand involved. If the wind is constant in both speed and direction, then the slip faces tend to be aligned and the dunes themselves move in tandem, amalgamating into much larger structures; maybe when the wind becomes more variable they become linear structures or ridges or whalebacks or dome-shaped, these latter without any slip faces at all. Some are small and manageable by the most timid of camels—two-humped Bactrians in Mongolia and on the vast expanse of the Taklamakan Desert in western China, single-humped dromedaries in the Sahara and Mewaris in the dunelands of Rajasthan and the Thar Desert of Pakistan. But in generally camel-less Namibia, dunes can be a thousand feet tall, and since the angle of repose of dry sand is 34° (rather steeper than the 28° angle of a pile of clover seed, less intense than the 45° of shredded coconut), the downslope can, at least for a motorist, be terrifying. Ralph Bagnold developed a fairly foolproof means of driving in such landscapes: accelerate smoothly and aggressively all the way up the stoss side of the dune, stop sharply at the summit with the vehicle's wheels hanging just over the very knife-edge at the top, let gravity pull the front of the car downward, and then, with a low gear engaged to act as a brake but without the driver touching the accelerator, let the car roll down the slip face at will before leveling out in time to clamber up the next dune's stoss face.

Bagnold did his early Tripolitanian* dune crossings in specially adapted Model T Fords; during the war his LRDG found eminently suitable fleets of Chevrolet WB thirty-hundredweight trucks, stripped of all unnecessaries and with wide, low-pressure sand tires. In Namibia we used long-wheelbase Land Rovers, and thanks to the competence of our Namibian driver, were never once stuck or bogged down in windblown sand.

All of our current knowledge of these wind-sculpted expanses of desert sand is, essentially, based on Bagnold's work in the mid-twentieth century; and of all we now know, one salient fact remains, unrealized by Bagnold himself but accepted all too well today: Deserts are expanding, and winds are helping them do so. Dunes are being blown into vegetated margins, denying farmers cultivable land and chasing villagers into regions that still have water and a less scorching climate. The prevailing winds that blow across the Sahara push its sands ever farther southward and westward, with red sand dust even staining the Atlantic Ocean and causing problems for navigators passing through its ochre mists. It is not a wholly new phenomenon; Lord Curzon had occasion to deem a wind that blows across the flat sandy plans of central Iran "the most vile and abominable in the universe," seeming to roll the desert out like a carpet, blowing for a hundred twenty

* Through its long and complex history Libya has been divided into three administrative regions—Tripolitania in the northwest; Cyrenaica in the east, bordering Egypt; and Fezzan in the south, petering out into the dune-filled emptiness of Tuareg country and the scattered oases of Chad and sub-Saharan Africa.

days straight each summer and burying all the fertile land that stands in its way.

Of desert expansion—desertification, as it has been known since the late 1960s—Bagnold knew little or nothing. He died in 1990, when global warming and climate change—and desertification—were words and phrases unfamiliar beyond the technical literature. But now such climatic developments are real, and, barring the attitudes of a few who would deny reality, are widely accepted to be so. And as temperatures increase around the world and weather patterns change, so in most of the planet's sand deserts that were themselves born of the wind, the very same forces that gave them birth are now playing further havoc with what they made, vastly expanding their extent and thereby placing millions of neighboring peoples at an ever-growing risk. Those who in the Global North profess anxiety about climate change think mainly of rising sea levels and ever more violent storms; we tend to forget that to many in the Global South, encroaching sands are every bit as grave a danger, and one that is considerably exacerbated by even the most gentle of winds.

{ 11 }

The closing chapter in Bagnold's book, which was placed there very evidently as an afterthought, is titled "Singing Sands"—the phrase in quotation marks, as the phenomenon of music-making dunes, described here a few pages back, seems somehow to have evaded the kind of rigorous scientific scrutiny found in Bagnold's earlier pages, as if it were more a fancy than a fact. But the connection between wind and music—whether involving sand or chimes or wires of platinum—is an even richer lode to mine, most particularly if the wind is not generated by the external forces of nature but by us, by humankind ourselves.

When we fill our lungs and then our cheeks and blow, we

can produce, if for only a little while, an airstream of thirty or forty miles per hour. Properly managed and directed through an instrument, this moving air, this *wind*, can produce music too—music every bit as sublime, and many would suggest rather more so, than atmospheric wind alone, blowing through trees or across the ocean, or along a dune or beach of singing sand. Wind instruments, as old as human history, are central to all cultures everywhere.

Chambers 20th Century Dictionary, my daily word catalogue of choice, devotes a full column to those hundred or so English words that sport the prefix *aero*, from *aerobatic* to *aerotropism*. Buried within the list, sandwiched between *aerophobic* (a fear of flying, or of drafts) and *aerophyte* (a kind of oxygen-loving plant), is the rather less-then-familiar word *aerophone*—which *Chambers* dismisses with a haughty economy, quite unworthy of the lexicographers' art, as *any wind instrument.*

The *OED*, as one might expect, rises more keenly to the task of explaining what an aerophone is: in its first, obsolete instance it was a Victorian musical instrument somewhat akin to a concertina; then Thomas Edison chose the word as the preferred term for a still imaginary invention that Alexander Graham Bell, who eventually did invent it, later called the telephone; and finally the *OED* gives us the principal meaning of the word today, an expansion on *Chambers*: *Any musical instrument in which the sound is produced by a vibrating column of air.* It lists six illustrative quotations dating from 1927 to 2005, the most recent offering as an example of the aerophone the Australian aboriginal device known as the *didj*—most would call it the *didjeridoo*—which the author claims is *among the oldest musical instruments on earth.*

These two entries—together with one marginally more substantial from the *Oxford Companion to Music*—are modest in scale. They all pale into insignificance when compared to what the mightily respected, 150-year-old *Grove Dictionary of Music* has to say about aerophones—for instead of a line or a modest couple of paragraphs, *Grove* offers up three full pages and a three-column appendix, reminding us that aerophones are one of the four great classes of musical instruments.* There are a great many of such aerophonic devices, since for centuries past humans have greatly enjoyed propelling air across instruments made in a bewildering variety of shapes and configurations, and so making an equally bewildering range of sounds with them.

And so—beginning with the most primitive aerophone of all, a single blade of grass held between two thumbs and then blown upon—a catalogue unfolds. Here follow a few aerophone examples, in no particular order: A recorder. A motor horn. A bosun's whistle. A bagpipe. A mouth organ. A siren. A bullroarer. A ventilating fan. A pop gun. A panpipe. A flute. A kazoo. A Swanee whistle. An ocarina. An aulos. A crumhorn. A saxophone. A trumpet. A cornet. A bugle. A conch shell. An alphorn. A didjeridoo. A trombone. A kurs. Uilleann pipes. A cornetto. A key bugle. A karafa (from Brazil). A bafiore (from the Lower Congo). A chanter. A serpent. A shawm. A pommard. A bassoon. A piccolo. A clarinet. A cor anglais. A tuba. A pellitone. A diapason. A bombard. A duduk. A piffero.

And, most complex and spiritually rewarding and emblematic of all: the oboe. To a layperson, a nonmusician, this most elegant of instruments represents an apotheosis, a congregation of human excellence and inventiveness that in the hands of some few players may result in a performance ethereally unforgettable. The oboe is an instrument that brings joy even to the melan-

---

* The others are *idiophones*, which are percussion-type instruments in which the very material from which they are made—as with a cymbal, a triangle, a gong—will create a musical sound; *membranophones*, which require a stretched skin or similar substance—as in a snare drum, a tabla, a taiko—that when struck will do much the same; and *chordophones*, musical instruments with strings—the cello, the harp, the shamisen.

choly. And it does so in spite of all attendant difficulties of its mastery, of which Ogden Nash offered a characteristically pithy summary: *The oboe's a horn made of wood. I'd play you a tune if I could. But the reeds are a pain. And the fingering's insane. It's the ill wind that no one plays good.*

But when they do, soft magic ensues. To hear one of the great pieces by Mozart, Tchaikovsky, Bach, or Albinoni played by one of the recent maestros, Heinz Holliger, Albrecht Mayer, or François Leleux, is to experience something approaching the sacred. And then, towering over all, Léon Goossens, born in Liverpool but the child of at least three generations of Belgian musicians of great distinction, who during numberless performances in the middle of the last century can fairly be said to have brought the oboe fully into the public consciousness, in Britain at least. This kindly, generous man, well-nigh universally thought of as the finest oboist in the world, became especially beloved during the worst days of the London wartime blitz, when he would play his oboe for the frightened homeless who were taking nighttime shelter from the bombs down in the stations of the London tube. It is said that when, thirty years later, he gave a public concert in Cornwall, elderly Londoners in his audience wept at their memory of the comfort he brought to them, playing sweet music on a wind instrument from which he could conjure such bliss and serenity in those bleak predawn moments before the sirens up above sounded the All Clear.

But how does an oboist, of whatever caliber, do such a thing? How does the passage of human-propelled air create such sounds in the first place? The technical explanations are fairly straightforward.

It all has to do with the manipulation of moving air. Central to all woodwind instruments is the plain physical fact of what exactly happens when blown air, *manipulated in a certain way,* acts upon the column of air that already exists inside an open-ended tube whose effective length can itself be manipulated by opening and closing holes, singly or in combinations, between one end and the other.

Each of the means of air manipulation—and there is a third, to be outlined a little later—is crucially important to the sound that will be emitted. The first, which can only be described as more than simply crucial, indeed *fundamental*, to the process, is the key to what makes a woodwind instrument a woodwind instrument—and that is the entity that divides into two, or actually splits, the air blown through the player's lips, and splits it in such a way that the two now smaller airstreams vibrate against each other and then in turn set up pressure waves within the still air that is already inside the instrument's playing tube. It is the frequency of the waves in this column of air that creates the sound; and that frequency can be altered, up or down, faster or slower, with the resulting note higher or lower, by opening or closing the various keys along the tube, thus making it a longer or shorter journey for the pressure waves and so altering the speed with which they pass along—altering their frequency, in other words—as required by the musical score that is presented on the stand before the player.

In a flute, the plainest in the woodwind family, the player's blown-out air is split quite simply by the sharp edge against which the player places his or her lips—some of the air goes upward, some downward, and as the two parallel airstreams progress along the flautist's tube, they are found to be ever so slightly out of sync because the distances involved in their passages across the sharp edge are microscopically different. A vibration begins, so the still air inside the flute starts to jostle itself in concert with the newly introduced vibrating air, and thus is music ready to be made.

However, the potential for air manipulation across a single sharp edge is necessarily somewhat limited. Which is why other, rather more versatile wind instruments allow the user to vary the inbound airstream by employing in the mouthpiece a *reed*—this being the truly fundamental component of most wind instruments. The player will direct his or her airstream by way of a mouthpiece onto the reed, which will then oblige by performing the air splitting and set up the subsequent vibration of the tube's internal air column—and all of this as a direct result of blowing across the reed's freshly cut and chamfered tip.

Reeds have been used to make music for centuries. The Ancient Greek story of Pan's lustful and goatish pursuit of the wood nymph and perpetual virgin Syrinx, for instance, places the humble reed at the very center of popular myth—since the kindly river nymphs helped the frightened and fleeing Syrinx to elude Pan by turning her into a hollow river reed, the better to disguise her. But Pan, no fool, saw through the ruse, promptly cut down the reeds, and tied a cluster of them together to make a set of graduated pipes from which he would conjure haunting music in his passion's memory: *panpipes* were an early musical instrument, and they were made—still are made—from the hollow river reed *Arundo donax*, from which almost all musical reeds are made today.

Arundo donax is a tall perennial and grows in temperate zones almost everywhere, being a classic invasive species from the Middle East—though rather more welcome than the tumbleweed, seen in the United States as a much-reviled botanical *arriviste*. *A. donax* can be used as a fuel or on construction sites (where it substitutes for the similar-looking bamboo), or it can be harvested for the making of musical reeds, to be blown across, and specifically so, by human agency. The very best-quality—meaning the straightest, the most perfectly cylindrical, with an outer skin that is thin and yet amenable to being shaved and whittled and scraped—*A. donax* grows in abundance in that coastal region of southern France lying to the east of Marseille, in the river valleys north of Saint-Tropez and Toulon. The musical reed industry, small but vital for the art form, is dominated by ancient family businesses in this *département*, which is otherwise far better known for its lavender and soap and delicate rosé wines.

And for a violent seasonal wind to which we will come later on, the *mistral.*

A player blowing through a mouthpiece reed will create pressure waves in the tube of a clarinet or a saxophone or in the very lengthy tube of a bassoon, or in that of its more modestly proportioned cousin, the oboe. The main difference within this family of woodwind instruments is that in the first two there is just a single reed, while in the bassoon and the oboe the reed is folded over on itself and then razor-split to create two sides, or blades, fastened together with a tightly wound wire, in effect doubling the reed.* The process of making a reed is exceedingly complicated, and most skilled oboe players tend to fashion their own, with the same attention to detail and single-minded obstinacy as those who make their own lures for fly-fishing. To perfect the two blades of a double reed there is much soaking and slitting and scraping and feathering and resoaking, the result being a reed that is peculiar and personal to that player and allows him or her to bring a unique combination of tone and timbre and pitch to the resulting notes of the instrument, giving each player an individual and immediately recognizable style.

The physics of what exactly takes place inside the silver tube mouthpiece from which the elements of a double reed protrude, with the two-inch-long device (which can produce its own sound, a croaking crow-like noise) then plugged into the upper end of an oboe tube just as the player begins to play, can be daunting. *Grove*, however, is succinct: *When blown the elliptical gap between the ends of the blades opens and closes, giving bursts of energy to the air column . . . the form and degree of "scrape" applied to the tip of the reed has a profound influence on its behaviour and sonority . . .*

---

* Those who manage to master the necessary skill to play instruments in this class guard their art jealously: the International Double Reed Society, based in a Baltimore suburb, protects and promotes their work with a near-Talmudic intensity.

Two variables alone—the shape and scrape of the double reed at the mouthpiece and the length of the tube in which the pressure waves pass along—might seem enough to allow the user to manipulate the airstream through the oboe; the longer the tube, the lower the frequency of the waves and therefore the lower the note that is produced. But there is a third variable, of no little additional importance: the manner in which the player arranges his or her lips and teeth and cheekbones and face more generally—the *embouchure,* as it has been known since the end of the eighteenth century, and for which the ever-reliable *Grove* offers an illustrative quotation: *The second octave is produced by a stronger pressure of wind and an alteration of embouchure.*

The instructive literature for the would-be oboe player is quite as daunting for suggesting the shape of the lips as it is for suggesting the shape of the reed. Very basically, the entirety of the reed—which, the manual warns, is very small—is placed directly onto the lips, with the top lip then placed under and around the upper teeth, to the alarm of most dentists, who regard the buccal gymnastics involved in playing the oboe well as placing the teeth in mortal danger.* The tongue has to be organized in such a way as to control and accelerate the airstream, thereby guaranteeing a fuller and richer sound, alive with overtones.

Having said all of this—and perhaps being tempted to accept Ogden Nash's sardonic verdict on the inability of most to play the oboe at all, let alone well—this most beloved of instruments, with its trinity of trickiness, did reap an eventual reward. Whenever an orchestra is preparing to play and the various instrumentalists start the process of tuning their devices one to another, it is

---

* The legendary oboist Léon Goossens, mentioned above, lost almost all of his front teeth and badly damaged his lips in a traffic accident in 1962. He spent many years working to recover his skills, and biographers remark on his courage and tenacity in doing so. But his embouchure was significantly altered, and though he returned to teaching, he never fully regained his earlier playing abilities.

invariably the oboe, sited in the orchestra's dead center directly in front of the conductor, that commences the entire procedure. The audience hushes itself, the lights begin to dim, and from the oboe a lone and haunting crystal-clear note sounds out to quiet all who are there to listen. The note is a pure and resounding A, its frequency 440 Hz exactly, and there is then a flurry of activity as those who wield string instruments tighten and loosen their various assemblages of gut and fine wire—and steadily, as each acquires and sets their note correctly and accurately, so all sounds fade away and leave just the one still impeccably warm note of the oboe, which echoes softly around the hall until it too is silenced, whereupon the conductor raps his baton, everyone pays attention, and the work begins.

Except in one fairly new musical circumstance: the playing of Ennio Morricone's theme to the 1986 film *The Mission*, titled "Gabriel's Oboe." In this case the oboist, a young woman who, having tuned the orchestra to perfect unanimity, sits back for the briefest of intervals before the conductor turns to her, and she, facing the baton straight on as all oboists in all orchestras these days do, picks up her cherished black-and-silver instrument once more and begins to play for two and a half minutes what is, by general popular agreement, one of the most hauntingly beautiful solo pieces of modern music imaginable. This, wrote an admirer, must be "what heaven sounds like." And all played on an instrument that has essentially been conjured into unforgettable life by the movement of air, by an airstream, by the wind.

{ 12 }

The maximum sustained speed of this chapter's *winds of a gentle nature* is around 25 miles per hour. This allows me, though only just, to include here the singular event that took place at Kill Devil Hills, on the Outer Banks of North Carolina, shortly after breakfast on the freezing gray morning of Thursday, December 17, 1903. This was when Orville Wright made the very first controlled, powered flight of a heavier-than-air machine—and with his twelve-second flight, at an altitude of about eight feet along a stretch of seagrass meadowland some one hundred twenty feet long—and into a headwind of some 27 miles per hour—he and his brother, Wilbur, inaugurated the whole new world of modern aviation.

THE WRIGHT BROTHERS were indeed sedulous and meticulous designers, and for years before their first flights, had calculated in detail the likely degrees of lift and drag that would be provided by the different shapes of the flight surfaces of their various machines. And in almost all cases their early calculations were derived from close observations of the behavior of one of the earliest recreational devices made specifically to harness the blowing of the wind: the kite.

A good thirteen years before that historic flight near Kitty Hawk, Sir Hiram Maxim—he of the infamously deadly rotating machine gun*—had won a patent for what the Wright Brothers would eventually make work. His application of 1890 noted *my invention is chiefly designed to provide for the construction of an aëronautic machine which can, while moving forward in the air, be caused to rise or descend at any desired velocity or to travel at any predetermined height above the ground. . . . I provide an adjustable covered framework or kite of very large dimensions. . . . For convenience of description I will hereinafter term this covered framework or kite an "aëroplane."*

So it can be argued that, unwittingly, the mechanics of the

* Hilaire Belloc used this awful weapon in his 1896 satirical chapbook, finishing a ditty about a "native" rebellion with the lines *Whatever happens, we have got / The Maxim Gun, and they have not.*

humble kite, a playful thing—indeed, a toy—would eventually lead humankind into a world of much more serious matters and manufactures, from jumbo jets to supersonic fighter planes, from Starlifter cargo carriers to humble crop dusters. But similarly humble though the kites may be, they are very old and time-tested, and invariably they provided those who flew them and fly them still in winds gentle and fickle and in between with an unparalleled amount of good, plain *fun.*

The image is an enduring one, seen hundreds of times down the centuries. There is a child—a boy, usually—on a meadow, or better still on a sandy beach with small white horses on the distant waves, and the sky is eggshell blue and chubby white clouds are scudding along it, and between a pair of these clouds is a small, gaily colored kite, diamond-shaped or pear-shaped or an isosceles triangle with its wooden cross members just visible at this distance, and from the point at its lower end comes a pendulous curving tail with maybe small cross tags of colored fabric flying in the breeze to give the kite some stability, with the whole flying affair connected to the boy below with a long, quite taut string that ends in a kind of homemade rudder-like handle that allows the boy, with subtle twists and turns and pulls and releases, to make the kite above him go this way or that or rise and fall or soar and sweep through the gusts and calms and the thermals up high, and beside the boy is a cluster of awestruck youngsters—cheering schoolfriends most probably—with, notably, a young girl that the kite-flying boy is trying to impress with his skill, and then to complete the picture there is the family dog, or maybe one or two others, and they are each looking up, as is everyone else in the picture, the dogs yapping at the mystery of an object in the near sky being yanked this way and that at the command—made with a very long version of a leash, no less—of the small boy here some hundreds of feet below, secure on the sand or the grass, taking advantage of the wind above to propel his toy toward the heavens. What joy! What amazement! What a sheer, unalloyed burst of fresh-air pleasure!

China, not surprisingly given its abundance of bamboo (for lightweight struts) and silk (for durable flying surfaces) and, especially in the Huang He River valley and the grasslands of Mongolia, vast expanses of breezy flatlands, lays claim to the kite's origin myth, with readily available images dating from the sixth century BCE. Japan got in on the act soon thereafter, and there are almost equally old drawings of contemplative-looking women, formally dressed in kimonos and with delicately held strings leading up to small, square-shaped mulberry-paper kites dancing in the sky. Kite flying subsequently became firmly annealed onto the Japanese cultural landscape, and during the May children's festival there are still well-attended gatherings featuring enormous kites weighing half a ton, looking anything but aerodynamic and requiring scores of strong young men hauling ropes to launch the devices off the ground. Depending on the wind speed—the ideal is between ten and twenty knots—a kite can remain aloft for several hours, six in one celebrated instance.

Pleasure is not the only intention in kite flying, joy not the only consequence. Perhaps the best-known image of a kite in flight is that painted in Philadelphia in the early nineteenth century by Benjamin West of his friend Benjamin Franklin performing his famous experiment to divine the existence of electricity in the air during a thunderstorm. It may seem wholly obvious today that lightning is a form of electrical discharge, but in 1752, when Franklin decided to satisfy his curiosity on the matter, it wasn't at all a certainty; the great man decided to use a kite to settle the question.

He had originally planned to clamber to the top of a local steeple to reach into the storm, but considering the risk involved to a forty-six-year-old man, saddled with a passel of homemade equipment, he made the late-day decision to behave more prudently and insulate himself somewhat by flying a kite into the first available patch of bad weather. Accordingly, on a sultry afternoon in June of that year he and his son William watched carefully the condition of that day's western sky and held up wetted index fingers to check on the direction and strength of the gathering breeze.

They were rewarded swiftly. A wall of black cumulonimbus clouds and thunderheads welled up over the city's western suburbs and heavy drops of a warm summer rain began to splash down. The Franklins took off, running for a hundred yards or so to a nearby park where they could launch the kite. This was a small handmade job, just a bamboo crosspiece and a silk handkerchief to provide lift. But there were modifications: Franklin had fixed a thin metal strut to the upper end of the kite's diamond-shaped frame, and to the lower end he tied a long guide string, its upper, kite-side end made of hemp, its lower, Franklin-side end of silk. His thought was that a rain-soaked hemp twine might conduct electricity, if there was any to conduct, while the silk thread would not absorb water and so would act as an insulator, and he could grasp that one and employ it to maneuver the kite without any risk of electrocution. Most crucially of all, he had also attached a large iron house key to the hemp portion of the rope, supposing it would add some further proof of any electrical charge. (Dr. Franklin also took along a homemade Leyden jar, a simple device made of glass and tinfoil that could store an electrical charge, with a view to discharging it in his home laboratory for study later.)

After a few moments the predicted storm began to crash around the pair, and with the help of twenty-two-year-old William they ran around the little park and launched their fragile confection into the gathering breeze. It rose, fitfully at first, before being swept up by the air currents and dashed into the heart of the maelstrom, with thunder pealing and lightning crackling around it. The pair, holding the guide string—grasping the silk end firmly, and examining the rapidly dampening hemp portion respectfully, somewhat gingerly—took to a doorway to shelter from the now pelting rain. They assumed the mien of scientific detachment, watching to see what happened next.

Benjamin Franklin during a Philadelphia thunderstorm in June 1752, using a kite and an iron door key to demonstrate the electrical nature of lightning. Benjamin West's painting, now hanging in the city's main art museum, was completed half a century later.

According to the British chemist (and discoverer of oxygen) Joseph Priestley, who was to become a close friend of Franklin's, the pair had to wait, and wait, and wait—until, just as they were preparing to pack up in exasperation and head home, Franklin suddenly noticed that all the loose threads on the hemp string had suddenly started to stand proud and to wave about wildly, excitedly. Very carefully he reached for the metal key; Priestley takes up the narrative: *Struck with this promising appearance, he immediately presented his knucle [sic] to the key and (let the reader judge of the exquisite pleasure he must have felt at that moment) the discovery was complete. He perceived a very evident electric spark.*

The poor man had, in other words and in the interests of science, felt an electric shock—probably of the strength that a bull might feel had he wandered into the wire of an electric fence. Not at all pleasant. Yet he retained an admirable sangfroid, offering the sparking key to the conducting inlet terminal of the Leyden jar and so managing to fill the phial, as he later termed it, with a decently sized charge of what he called "this Electric

Fire." Whereupon the pair, more than amply satisfied with their afternoon's work, pulled down from the sky their now breeze-battered kite, stashed their various wires and jars and sodden clothing in bags and hurried home, eager to tell the world what they had just discovered.

Benjamin West painted his famous small portrait in 1805, more than half a century after the experiment that his artwork would then make famous, and he did so most probably to mark the centenary of Franklin's birth in 1706. (Franklin himself had died fifteen years earlier.) Would he have approved? West shows him as elderly, wrinkled, and with white hair, which at the time of the kite experiment he most certainly did not have. Old he may look, but with a heroic stance, his face directed in defiance toward the heavens, his robe blowing in the wind. There are four cherubs in various degrees of undress, two of them conducting experiments. Up above there are dark, menacing clouds that break to show distant lightning flashes. The hemp rope angles across the upper half of the frame, the carefully delineated house key dangling from it, the equally carefully delineated spark extending from the key to Franklin's index finger—an image of a digital encounter owing rather too much to the Sistine Chapel and Michelangelo. It is difficult to imagine as pragmatic and polymath a figure as Benjamin Franklin warming to a painting that was so overtly romantic, did not manage to feature any imagery of the kite itself, and was so very insignificantly small, less than half the size of the already quite puny *Mona Lisa.* (The comparison permits me to note that Leonardo da Vinci also drew kites—lots of them, in all shapes and sizes.) But even if Franklin might not have cared overmuch for the particular depiction, the US Postal Service liked it fine, and in 1956 *Benjamin Franklin Drawing Electricity from the Sky* by Benjamin West was featured on the three-cent first-class stamp, and anyone sending a letter anywhere within the United States back then could affix it to the envelope, as many millions of letter-writers did, further helping to making Franklin's bold experiment in 1752 one of the best-known illustrations of the scientific method of all time.

{ 13 }

Science and kites became still more intertwined in the nineteenth century, once users began to look more closely at why a gentle wind might lift and propel forward an object—a metal key, a part of a pagoda in the making, a lantern, a pet to be brought across a river, a man to be freed from his castle prison—and why at other times a kite might seem to fall backward, to drop unexpectedly, precipitously, to behave erratically. All of this might be great fun to the boy on the beach, and tugging the string this way and that and maybe running along the beach to get some kinetic energy up to the kite to prompt it to fly just a little higher might work to solve the problem—all of this kind of thing is and long has been what makes kite-flying so spirited a pastime. Even when, as Seamus Heaney famously noted in his 2010 poem "A Kite for Aibhin," there is a momentary catastrophe and the kite breaks loose from its string.

> Air from another life and time and place,
> Pale blue heavenly air is supporting
> A white wing beating high against the breeze,
>
> And yes, it is a kite! As when one afternoon
> All of us there trooped out
> Among the briar hedges and stripped thorn,
>
> I take my stand again, halt opposite
> Anahorish Hill to scan the blue,
> Back in that field to launch our long-tailed comet.

And now it hovers, tugs, veers, dives askew,
Lifts itself, goes with the wind until
It rises to loud cheers from us below.

Rises, and my hand is like a spindle
Unspooling, the kite a thin-stemmed flower
Climbing and carrying, carrying farther, higher

The longing in the breast and planted feet
And gazing face and heart of the kite flier
Until string breaks and—separate, elate—

The kite takes off, itself alone, a windfall.

{ 14 }

At the close of the nineteenth century, as they were approaching success with their long-held notion that they might be able to build a true engine-powered and controllable heavier-than-air flying machine, the Wright brothers looked to kites, quite specifically, for much of their aeronautical inspiration. They already knew the basics. They had, for example, read the famous paper written in 1809 by Sir George Cayley, who—wealthy Yorkshire landowner with a keen interest in the drainage of his meadows though he was—was obsessively fascinated by flight, and can fairly be said to have come up with the mathematics to prove it, the idea of the *airfoil.* His reputation as the "father of aviation" derives from his realization of the four forces that operate on any object attempting to fly through the air—weight, lift, drag, and thrust. He also drew diagrams showing the simple logic behind why an airfoil, if perfectly constructed and moved in a very specific manner, has the unopposable need to lift itself upward in the air. A slender wing inside a moderate airstream, if its top and bottom surfaces are parallel and equally flat, will barely disrupt the flow of air. But if the upper surface—and only the upper surface—is bowed upward, if it is made with a *camber* to it, then the air flowing over it will necessarily have to travel a slightly longer distance than the air that flows—at the same speed, of course—over the flat and unbowed surface below. The air above will be marginally thinner, therefore. It will be at a lower pressure. And this will cause the wing to try to move up toward the area of lower pressure, just as air in general likes to move from areas of high pressure to those with lower—to create, as we are well aware by now, that movement of air we know as wind. It is much the same with a wing—the airfoil, being swept through the air at higher and higher speed, will reach a point where it is physically compelled to raise itself up and take whatever is connected to the wing up with it.

You can feel this phenomenon—you can *get it*, the very basics of aerodynamic lift and so the very basis of aviation—if you are ever lucky enough to be invited to sit in the cockpit of a commercial jetliner at takeoff. You will have been assigned a runway that offers you wind blowing as directly toward you as possible. You know how much fuel you are carrying, how many souls are aboard, how heavy the aircraft is on that particular day, how long the runway is, and what speed you must achieve such that the differential of the pressure above and below the wings, above and below the airfoils, lifts the aircraft into the sky. You are cleared by the control tower. You spool up your engines to full takeoff power, their thunder sounding even as far forward as the cockpit. You accelerate smoothly down the runway. Your colleague, watching the airspeed indicator, calls out *V1*—the speed above which you must now take off, whatever your situation. The point of no return. Seconds later you call out *Rotate* and ease back the stick to raise your aircraft's nose—and then you feel it, the irresistible upward tug of thousands of horses pulling you up and into a new altitudinal airstream, unleashing you, your scores of tons of airframe metal, your engines, your passengers, your cargo, all that is in normal circumstances held firmly anchored to the ground by the insuperable force of gravitational attraction, and now freed from "the surly bonds of Earth" to reach up and, as John Gillespie Magee Jr.'s famous wartime poem "High Flight" has it, to touch "the face of God."

Such, maybe somewhat fancifully, were the distant dreams of Orville and Wilbur Wright, and it was their patient, yearslong study of flight surfaces—airfoils—within the specific world of kites flying in winds both gentle and unpredictable, that would lead the brothers to their eventual hard-won triumph. And control was really what the kites most notably offered to them: how to control a plane in flight.

It is all very well to suppose that flying an aircraft is little more than allowing those thousand irresistible horses to pull you up into the sky and then cavort gaily across the heavens. For years past, any flying student was offered the axiom most emphatically on the first day up in the air that *takeoff is voluntary, but landing is compulsory.* And in order to land in one piece, so the pilot of an aircraft, whether a shuddering Kitty Hawk biplane with wings made of canvas and beechwood coming down to land on grass in a ten-knot headwind or an F-35 fighter landing on an aircraft carrier in a Force Ten gale, has to have control of his machine. And back in the late nineteenth century, such proved a challenge of quite fantastic difficulty.

But the Wrights' close study of the behavior of kites taught them how control of a powered heavier-than-air machine might actually be achieved, one step after another. There are basically three ways in which a plane can be moved away from its initially intended trajectory: it can roll, with one wing moving up and the other down; it can pitch, with the nose moving up and the tail pushing down; or else it can yaw, with the fuselage moving sideways, leftward or rightward, around a vertical axis. Time after time, early experiments showed how powerful these three forces can be: a plane might take off, its nose might suddenly pitch downward because of a sudden gust of wind or an error by the pilot, and, if it proves impossible to bring the nose back upward, then the hapless aircraft plunges nose-first into the dirt, with disagreeable results. Likewise its starboard wing may slew downward, the pilot is unable to bring it back level, and the wingtip gouges a scar in the meadow, destroying itself and ending the flight in a confusion of wrecked spars and torn cloth. Or else the pilot tries frantically to keep the plane from turning suddenly off to the left and toward that distant line of trees until it gets tangled up in a mess of leaves and branches and the engine stops and fire erupts from a broken fuel line. These and a hundred other eventualities showed that humankind, however ambitious in its plans, as Hiram Maxim had laid out, to conquer the skies, was proving quite incapable of taming the machine in which it might do so. Or of achieving what is called three-axis control, in which you pitch, roll, or yaw only when you want to, not when the plane decides to do so on its own.

The Wrights eventually solved all this. They built a small biplane kite in 1899 and attached four wires to each of its wings; and, while allowing it to soar and dip and rise in the breezes that blew in from the Atlantic Ocean, they learned how to warp the wings, how to change their shape to change the airflow across their upper surfaces, and so decide down on the ground how to roll the flying kite in this direction or that as they saw fit. That took care of the roll; using similar techniques they managed to keep the kite from turning off course and to keep its plane-shaped nose from heading upward or downward. In sum, they managed to achieve with wires that essential degree of three-axis control that is now achieved with the use of flaps, with elevators and ailerons and trim tabs, and with rudders.

Their short, successful first powered flight on that cold December day in 1903 at Kill Devil Hills was no fluke. They had worked long and hard at the theory of the thing before they swung that laminated spruce propellor and so cranked their little engine to life; by the time they were ready to lift off they knew they had control, and it was their tireless preparation using time-tested devices that had been named a long while ago for a soaring bird of prey they hoped one day to emulate that brought them the success, and the fame, and all the benefits and trials of today and the vast empire of aviation.

{ 15 }

A hiker making his or her way uphill from Honolulu or Waikiki or Diamond Head or the university campus at Manoa, passing from the lowlands of southern Oahu up toward the great volcanic ridge at this Hawaiian island's spine, will eventually, if fit and strong and tenacious enough, reach the summit lookout of Mount Tantalus. Students down in Honolulu named the mountain back in the 1840s for the Greek god who was tortured by the presence of food that was always out of his reach and wine he was never able to drink—the summit similarly always seeming to elude the would-be climber. Elusiveness aside, what really challenged the hiker about Mount Tantalus was the wind. Over the final few yards what until now has blown soft and warm and moist with tropical humidity starts to pick up in strength until, at the summit itself, it hits the hiker with eye-watering force, slamming into you with all the power of working trade winds howling directly up from the North Pacific Ocean and blowing here at full tilt. The higher summits all around are thick with wind-driven swirls of wet mist; ahead to the north and east are the sun-swept beaches and military airstrips and bungaloid housing developments of Kailua and Kaneohe. The foliage up on the hilltops is thick and resplendent and aromatic—guava and staghorn fern and cassias and monkey pod and myrtle, and with giant eucalyptus and fast-growing acacia koa trees punctuating the skyline. It is high country of a peerless beauty, in sheltered nooks a quiet world miles away from the frantic beat of tourist Hawaii, all peace and birdsong and endless perfumed breezes.

It was to such a landscape that Howard and Betty Liljestrand were inexorably drawn back in the 1940s. They were in the medical profession, he a doctor, she a cytologist, he from New York State, she from Iowa. He had had an unusual life thus far. His father had been a medical missionary in western China, and the young Howard, who had been first taken to Chengdu when he was a four-year-old, thought long into his adulthood of China as home, and when he returned to the United States to pursue his own medical degree, following in his father's path, supposed that he would return there in due course and remain contentedly there for the rest of his days. But as so often, *events.* Happenings—civil strife in China, mainly—intervened, and a now married Howard and Betty (they met while he was at Harvard and she at Columbia, both doing summertime duty at the great marine biology laboratory in Woods Hole, Massachusetts, with Betty finding herself one microscope short and Howard just happening to be minding the equipment storage room . . .) moved to Hawaii, essentially to wait for China to settle itself down and allow his return. But that never happened. Eventually, after years of hemming and hawing, the pair decided to remain in Hawaii for good, and to build themselves a house.

They had no wish to settle by the ocean, however; Howard's fondest childhood memories were of the family summers in China, their bungalow in the cool, high hills around the Sichuan furnace cities on the plains. The house they would now build for themselves in Hawaii would be on the slopes of Mount Tantalus, and they would harness the wind up there to make it as cool and pleasant a home as those old bungalows had been, high above Chengdu.

They turned to an architect whose background was as unusual and as exotic as theirs had been. He was Vladimir Ossipoff, and was by the time they met among the most prominent members of the architectural community in all Hawaii. He had been born in Vladivostok in Imperial Russia early in the twentieth century, his father a soldier who went on to become military attaché in the Russian embassy in Tokyo. However, events: the Bolshevik Revolution of 1917 took place, effectively stranding the Ossipoff family in Japan—until the 1923 Great Kanto earthquake frightened Mrs. Ossipoff and drove them away altogether, with young Val Ossipoff—by now fluent in Russian, Japanese, and English—finding himself in Berkeley, California, to complete his education.

One prior encounter seems to have sealed his fate: just before leaving Tokyo he had seen and been mightily impressed by the start of construction of Frank Lloyd Wright's Second Imperial Hotel (the first having been burned to a crisp). As a direct consequence of this epiphany, Ossipoff would choose architecture for his studies at the University of California's Berkeley campus, and then choose it as his profession when he acquired his degree. In 1931 he moved to Honolulu, almost on a whim, finding two features of the islands that would keep him there for the next sixty-seven years, the rest of his life. His first realization was that the architectural landscape of Oahu, when he first arrived, was dismal in the extreme, with the capital in particular little more than a confused mess of the hastily thrown-up and the decidedly jerry-built, and which he vowed there and then he would try to improve. The second was born of the very evident swelling of the ranks of incoming residents—professionals, eager to hang out their shingles in so spectacularly lovely a place, who had available funds and imagination and a keen desire to live beautifully amid the natural beauty with which all the Hawaiian islands were generously endowed. After conducting for his initial years on the islands his "war on ugliness," Ossipoff then moved on to become the renowned champion of the so-called *kama'aina* school of architecture, building Hawaiian houses for Hawaiian residents—and, in as many cases as possible, working sympathetically with the natural phenomena, the cool and perfumed winds very much included, with which the islands were blessed.

The bond to be struck between Val Ossipoff and the Liljestrands was therefore more or less inevitable. They duly met, the Liljestrands outlined their ideas, and they walked Ossipoff through their recently purchased acreage a thousand feet up on the southern slopes of Tantalus. Their plot was potentially quite magnificent, a narrow half acre terrace perched at the tip of a slender ridge between two steep, lush valleys on the leeward side of the Tantalus slopes and approached by the serpentine curves of Tantalus Road, about three hundred feet below the summit.

It took some few years before the Liljestrands committed and took the step for a project they feared might keep them in debt for the rest of their lives, but in 1948 the excavators and draglines and tractors moved in and trees were felled and thousands of tons of rich red volcanic earth were scraped away, revealing from the newly cleared lot perhaps the best of all possible views of Oahu—from the familiar peak of Diamond Head on the left to Waikiki and the Punchbowl Crater and downtown Honolulu directly ahead to the battleship-busy lochs and gantries of Pearl Harbor to the right, with the endless hammered-pewter sheet of the Pacific rolling out to the blue horizon as backdrop to it all, blending seamlessly with the eggshell blue of the Hawaiian sky. The building of the house got underway, and it took two years. And though Betty Liljestrand was punctilious and demanding (especially of the kitchen arrangements), and even though Ossipoff himself insisted that the furniture, much of it built to his own design, be arranged in particular and immobile ways, and sometimes oddly—the bed in the master bedroom was sited at a curiously nonintuitive angle, for example—there were no major fights as there so often are with such projects, and the two sets of families, the Ossipoffs and the Liljestrands, unusual by background and temperament, became firm and lasting friends for the rest of their lives.

The house was California redwood up above, with detailed mortices and tenons and pegs and joints painstakingly assembled by imported Japanese shrine builders, concrete in forms polished and rough as the room demanded beneath, mid-century modern styles within and with the overhanging roof sheathed in metal—this last a nod to the great Ceylon architect Geoffrey Bawa, who once famously remarked that the all-sheltering umbrella was in many ways the most perfect architectural form, and who had a keen appreciation of how wind and weather might best be incorporated into structures created in the tropics.

Val Ossipoff took with utmost seriousness the matter of how the cool Hawaiian northeasterly winds, the classical *trades*, might work with the house he was now contracted to design. And though the Liljestrand House, finished in the mid-1950s

In designing Hawaii's Liljestrand House in 1952, the architect Vladimir Ossipoff sited the property very precisely to catch the prevailing northeasterly trade winds and then the Venturi effect to channel these winds through the property such that even in midsummer a cool wafting made air-conditioning quite unnecessary.

and providing a legendarily comfortable home for the eventual family of six, is regarded a nonpareil of stylist integrity for all who celebrate the charm of mid-century modern, its less obvious contribution to the magic, the best architectural detail of all, lies in the manner in which Ossipoff dealt with these winds.

He employed the principles of the Venturi effect, a phenomenon known since its discovery by the eighteenth-century Italian physicist after whom it is named, in which air forced through a narrow opening both speeds up and loses its pressure. Ossipoff—aware that such principles of physics lay behind the design of some of the ancient Mughal palaces in the faraway red-hot cities of Rajasthan—opted to try such an arrangement here in Hawaii. Here at this site on the leeward side of Mount Tantalus his clients could expect a formidable aeolian cocktail: cool trade winds spilling over the ridge from the northeast, and occasional violent updrafts of hot air from the Waikiki side shrouding the ridges and ravines all around. Using the Venturi effect might not tame the updrafts nor limit the rainfall, but it could make use of the trade winds to keep the house pleasantly cool all year round, and without recourse to any unseemly electromechanical means of conditioning the air.

The house that he built is entered from what is technically its rearmost side—the main door and the vestibule and the carport are what one sees first, at the end of the long jungle-covered driveway. This side of the building is angled directly toward the northeast, the windward side, and when the trade winds are blowing they hit this side of the building and, beside the doors and the carports and the other quite ordinary features of a fairly modern house, pass into a series of adjustable venturis. The wind blows into the house through a mesh of small openings in their redwood screens and, just as predicted by the Italian mathematics of two centuries before, the airstreams from outside speed up as they pass through the narrow openings and emerge at the far side, now being fully inside the vestibule where visitors might leave their coats and shoes. Here the reasonably sedate wind outside would be accelerated to a startlingly faster velocity, almost making it uncomfortable to linger in this part of the house. But as Ossipoff reasoned, nobody tends to linger in the vestibule of any house; no one stays in the hallway or the mudroom or the air lock—and in any case, the size of the venturis' openings could be adjusted and made larger or smaller to help moderate the speed of the inflowing air, just in case any eccentric or perverse visitor decided they did in fact want to hang around in the cold little gale for a while.

But then this faster-moving air spreads itself into the more cavernous rooms on the leeward side of house, into the living rooms and bedrooms with their capacious interiors and, most crucially for the fluid dynamics of the situation, their very large windows—windows that do rather more than simply provide the spectacular panopticon of Honolulu views, though they certainly do that. What the larger windows otherwise achieve, based on Signor Venturi's physics, is to encourage the incoming air to slow itself down steadily as it passes through the rooms and moves from the windward to the leeward side of what would become Ossipoff's best-known and most fondly regarded creation. And in passing as they do, the trade winds are first speeded and then moderated, transmuted in this adroit evolution into wafts of what seems a permanent cool breeze wherever the occupants—family or friends or, in more recent years, those members of the public who come to witness and pay homage to the architect—might care to gather. The airflow is never so strong as to flutter a newspaper or nudge away a coil of incense or a hint of perfume, yet it is always there, keeping the interior of the house imbued with a very real sense, a feeling on the skin, of an ozone-hinted freshness and delight. Though it is absurd to suppose one could ever truly tame the wind, it's clear in Val Ossipoff's immensely thoughtful making of the Liljestrand House that something on a lesser scale can be achieved. With care, humility, respect for nature, and attention to detail one can perhaps soothe the wind somewhat, and perhaps turn it briefly to fleeting human advantage. And even in doing so, the wind passes from the mountain down to the sea almost wholly uninterrupted—its passing via the Liljestrand's home just the briefest of diversions—as winds should always be.

{ 16 }

And then, at the coastline down at the base of the mountain, there are breezes, gentle winds—similar in strength to those that move thistledown and tumbleweed, radioactive particles and light-shrouding smoke, that make oboes sing and kites fly and keep well-made structures cool—but the coast is where gentle winds also gain muscle and energy and potential. At the shore, some kind of winds regularly—every night and every day, in some places—become strong enough to move more than gossamer and mulberry paper and become able to lift giant things, shift ships and people—and make power. These are the winds that do work, and they have had and always will have the ability to change the face of our planet and the lives of those millions who live upon it. These are the working winds, the winds that take us places from one part of the world to all others.

There probably was a time when early man—pre-Aristotelian man, that is—took the gentler winds more or less for granted. Providing these breezes were benign, it was perhaps prudent to leave their mysteries be and not to inquire too deeply, lest the gods behind them and who gifted them to us mortals become somehow offended. But once winds started to be harnessed to perform work, things began to change in the relationship between man and the wind that he now wished to be at his service. Those who made use of the potentially limitless power of the wind would learn to adjust their harnessing of its potential so as to take ever greater advantage of this strange and initially inexplicable force.

What was the force, they now needed to know. What exactly was the wind, how did it happen, how could it be predicted, employed, tamed? Its full and proper use demanded that it be a mystery no longer.

CHAPTER TWO

Something in the Air

In all things of Nature, there is something of the marvelous.

—Mistakenly attributed to Aristotle, but encapsulates the sentiment expressed in *The History of Animals*, 343 BCE

For any wind to blow across any surface anywhere, two components are essential. In order to experience the phenomenon, any celestial body—be it a star, planet, moon, asteroid, comet, dwarf, or something even smaller—must first possess an atmosphere. And there must also be, either within the body itself or beyond and yet relatively close to it, a source of energy that from time to time will cause part of this atmosphere to increase or lower its temperature. And as a simple physical consequence of this heating or cooling, so to allow the molecules that comprise the atmosphere to move. To move upward and downward or, more usually and portentously, to move sideways. To blow. To become wind.

No atmosphere means, quite simply, no wind.* Our moon, for example, has no wind—precisely because it has essentially no

---

* The phenomenon known as the solar wind flows from the surface of the sun and out into the vacuum of outer space. Its ceaseless cascade of electrons and protons has a wide range of effects—it creates spectacular polar region auroras, it can do great damage to Earth's electrical grids, and over billions of years is believed to have wholly eroded the once merely fragile atmosphere of our neighbor planet

atmosphere; its gravitational field is not strong enough to keep any meaningful volume of air pinioned to its surface. And as it happened, this chronic condition of lunar windlessness posed something of a dilemma for the first American astronauts who landed on the moon in July 1969.

Not unnaturally, the American government wanted to memorialize their visit, and NASA's majestically named Committee on Symbolic Activities for the First Lunar Landing decided, on the advice of many august bodies, that Neil Armstrong and Buzz Aldrin, the pioneers, should plant a national flag close by their landing spot. But flags fly only when there is wind to help them do so—and in this windless vacuum the Stars and Stripes would just hang impotently, the very antithesis of the pride of American achievement that the committee had been established to demonstrate.

How to make a flag appear to fly where flying was not possible? Skilled NASA engineers promptly bent to the task. Just three days before the rocket carrying the *Apollo 11* capsule was due to blast off from Cape Canaveral, a slender steel tube weighing just nine pounds was attached by zip ties to the forward leg of the moon landing vehicle, just behind the ladder at the base of which Neil Armstrong would soon make his famous (and Committee-approved) "one small step for Man . . ." declaration before planting the first human boot in the hitherto untroubled lunar dust.

Inside the tube was a three-by-five-foot high-impact nylon Old Glory, designed to be large enough to be seen on television, together with two gold-anodized aluminum tubes. One of these was eight feet tall, to serve as the flagpole, and the other was five feet long and cunningly fashioned to run through a sleeve that had been sewn into the flag's upper edge. Once assembled—and footage of

---

Mars. But it does not operate on or within an atmosphere. It doesn't move air. It is well outside the scope of this book.

NASA designers made certain that the US flag planted on the lunar surface by astronaut Neil Armstrong in 1969 gave the appearance of flying in a light breeze. But since it has no atmosphere, the moon has no wind—the image a reminder of the interstellar ubiquity of PR.

the landing shows that it took the well-rehearsed pair no more than three minutes to get the whole thing locked together, planted, and upright—the arrangement looked very much as though the flag was flying in a gentle lunar breeze, with a slight fold of the fabric brought about by the engineers' previously mentioned cunning design. Armstrong and Aldrin then stepped back and saluted their flag, sealing the moment in history.

The windless scene was made to look moderately realistic, as though it might be yet another instance of American Manifest Destiny, of territorial expansion, as if the flag had been planted on some newly won acre of prairie—with just a light air stirring the fabric—somewhere deep in Nebraska Territory.*

* The monument turned out to be only temporary, however. Exhaust gases from the departing spacecraft—ironically the only prevailing wind the moon has ever

There are around three hundred recognizable moons in our solar system—the number keeps climbing, as satellite detection systems and telescope lenses become ever more sophisticated and perceptive. Currently Saturn has the most, with 146, and Jupiter is second with 95—although Jupiter's particular menagerie of lunar orbiters is evidently more easily seen from Earth, with the original four—Io, Europa, Ganymede, and Callisto—spotted as long ago as 1610, by Galileo.

Only a paltry five of these hundreds of bodies possess detectable atmospheres: the yellow-tinged Galilean moon Io, which sports a powerfully active peppering of some four hundred volcanoes, has a thin and patchy coating of sulfur dioxide gas; Titan, which flies some seven hundred thousand miles above Saturn, has a thick covering of nitrogen and methane; as does Triton, hovering happily high over Neptune. Until the Hubble Space Telescope began surveying in earnest following its repair in late 1993, it was assumed that these three—Io, Titan, and Triton—were the only atmospheric moons; but then a further two of the original Galilean quartet, Ganymede and Europa, were found by Hubble to possess slim layers of oxygen and nitrogen adhering to their bitter-cold surfaces, bringing the total to five.

All of the eight true planets in the solar system, the motherships to these flotillas of moons—as well as the sun herself, the star that is mothership to the solar system entire—possess serious atmospheres, robust gatherings of gases of various kinds and in various proportions that blanket themselves above the rocky or icy surfaces below. The sun, uniquely, is in essence a continuously running thermonuclear fusion reactor, with its atmosphere of hydrogen and helium being transformed, the former into the latter,

---

known—knocked the flag over and sent it tumbling away, and later visitors to the region have failed to find any remains. Subsequent lunar expeditions elsewhere have required their flags to be placed a more respectable distance from the takeoff sites.

at the rate of some six hundred million tons of gas each second, the endless explosions producing the heat and light energy for which the sun has long been cherished. The other planets are clothed in somewhat more benign fashion. Mercury has just a slender covering of a mixture of very hot oxygen, sodium, hydrogen, and helium; Venus, a viscous porridge of carbon dioxide and nitrogen; Mars, a respectably muscular layer of the same two gases; and then the big boys—Jupiter, Saturn, Uranus, and Neptune—playing host to clouds of hydrogen, helium, and, in the case of the two most distant, methane.

The simple presence of an atmosphere, though, fulfills only one of the necessary two requirements for the generation of wind. A heat source is needed as well. The most obvious source in our own solar system is of course the sun, at least for those planets and their moons that are relatively close by. As it happens, though, some of the most notable windstorms discovered in the system have tended to be in the more distant bodies, where the sun is just a tiny faraway object in the sky, its heat effect minimal. The energy sources that stir up the atmosphere and cause winds in those bodies tend to be volcanoes; the moons that perhaps demonstrate this best of all are Titan and Triton, each playing host to an array of spectacular weather.

Titan, for example, has windborne dust storms—big, rolling monsters, not unlike a Sudanese haboob, or such as you might see and try to outrun on the road between Marrakesh and Agadir when the sirocco is blowing sand up from the Sahara. The winds that blow Titan's sand off the moon's equatorial fields of desert dunes would seem to be generated by the temperature differentials within the atmosphere triggered by the methane rainstorms that are a summertime feature of this highly active body. Similarly, Neptune's Triton has observable winds, mostly fairly gentle and almost certainly triggered by atmospheric anomalies brought about by the eruptions of scores of recently found *cryovolcanoes,* intriguing deep-space versions of the lava volcanoes that are ranged across the Earth. On Triton the geologic process is little different: fluid under pressure wells up from within the moon's crust, finds a weak spot, and gushes out into the open, sending a plume of white smoke up into the sky. The difference is that on Earth the fluids from a volcano are hot and molten, whereas on Triton they are plastic and ice-cold. Nonetheless, the sudden presence of the intruder upsets the delicate temperature equilibrium of the local atmosphere, which then starts to shift and move about, up and down and sideways as Triton's wind.

Such atmospheric disruptions occur on a far larger scale on the planets themselves—all of which, with the possible exception of Mercury, enjoy or suffer tremendous amounts of wind. Neptune most of all: for reasons that still intrigue astronomers, gales of over 1,500 miles per hour regularly sweep across the entire planet, bitter cold and supersonic winds that are many times more powerful than any recorded or imagined earthly hurricane and blow uninterrupted by any mountain chain (Neptune is a billiard-ball-smooth body, composed mostly of frozen gas), and that are constantly fed by mysterious forces so that they seldom stop blowing at all.

A similar lack of surface features on the planet Jupiter, which at its closest can be a bit more than two million miles away from Neptune, is probably responsible for one notable feature of its unique weather system, first seen in the seventeenth century and still very much in action today: the Great Red Spot. This is an immense cyclonic storm, its diameter furious with swirling gales running at many hundreds of miles per hour, the whole system nearly twice as large as Earth itself. It is sited, looking much like an angry carbuncle, about 20 degrees south of the Jovian equator, and is by far the largest and most obvious of a number of excrescences that stand out against the bands of atmospheric activity that mark Jupiter as a most unusual planet. The Spot grows and fades over time, and since it was first seen in 1665 it has either changed position or was wrongly located by its first observers (whose telescopes in those days showed reversed images, causing much confusion).

As with Saturn's rings and the ultrafast backward-blowing pale-blue gales seen on Uranus—and, indeed, as with the frequent seasonal dust storms that are kicked up on Mars, events that are anticipated and planned for by those who would colonize this nearby and relatively benign neighboring planet, and events that are also dramatized by those who already make Hollywood films about living in such a place—all of the solar system's planets (save for Mercury, as lifeless as the moon, with almost no weather at all) seem to have just a single primary kind of recognized climatic feature. It may be supersonic winds, rings of colored dust, red spots, cryovolcanoes, or sandstorms, but there seems to be an absence in any one of the planets of the sheer variety of weather phenomena that are recognized on Earth—a fact that derives not so much from the detailed ways in which our home planet has been observed over the centuries but because of the daunting complexity of Earth itself, a complexity that makes it the fortuitously unique home for such a vast abundance of what the rest of the solar system lacks—and that, of course, is life.

THE BASIC PRINCIPLES involving the two components deemed essential for the generation of wind—an atmosphere and a source of heat energy—obtain across not just our solar system but the universe. They are universal—are, in essence, a physical constant. They are underpinned by the First Law of Thermodynamics, of blessed school days' memory, which according to its Victorian proponents holds that in a closed system energy can be neither created nor destroyed but only transferred as one kind of energy to another. Thermal energy—heat—can be transferred into kinetic energy—motion—and vice versa. Apply heat to any substance—solid, liquid, or gas—and the molecules of which the substance is comprised will immediately start to move, to jiggle, to vibrate, as the energy is transferred from one kind to another. The result of this energy transfer, when it comes to the heating of a solid, is that the tight, static, and structurally well-organized molecules that make up a solid will become agitated; the solid will begin to expand in size and, if yet more heat is applied, the internal organization of the molecules will become so disrupted that the expanded solid will lose its rigidity and solid integrity and will melt. It will change phase and become a liquid.

Much the same will happen with a liquid. Its molecules, which are already in much wilder and more random motion than they are in a solid, will become ever more lively and uncontrolled as more and more thermal energy is applied, until some of them break free of the liquid itself and leave the imprisoning confines of its surface. In so doing, the molecules that leave will start to deplete the remaining liquid, which is said to be evaporating—doing so more quickly as the temperature continues to rise.

Eventually the number of escaping molecules becomes so large that they have no need to bother with the trials of breaking free of the surface but escape within the liquid itself, turning into bubbles of gas in a process we generally regard as boiling. The liquid is by this time fully engaged in the process of altering its phase from liquid into gas, which it will continue to do as long as the heat continues to be applied and until all the liquid has, quite literally, boiled away.

This kind of thermodynamic process becomes specifically relevant to the production of wind when it deals with the phase that is already in the most agitated of the three—solid, liquid, and gaseous—when it exists as a gas at the ambient temperature of the system. An atmosphere is by definition a gas—"The spheroidal gaseous envelope surrounding any of the heavenly bodies" according to the *OED*—and it behaves when exposed to the arrival of thermal energy exactly as the First Law dictates: its molecules, already in the very motion that defines their gaseous state, begin to move ever more wildly and randomly, to move ever farther from one another as the effect of the heat causes the gas to expand. A cubic centimeter of a heated and expanded gas has, by definition, fewer molecules than its cooler brethren, and so has less mass. The concept of weight then enters the picture, with weight being the degree to which gravity acts on the mass of a body. Gas with low mass is therefore lighter than cooler and more dense gas that has a greater mass; and, as a consequence, the gravity that keeps heavier gases more firmly anchored to the planet's surface releases its grip on the warmer and thinner gas, allowing it to rise above its cooler kin and into the upper reaches of the atmosphere. Warm air, in other words, rises.

THIS NECESSARILY CUMBERSOME explanation—logical, mathematically demonstrable, and quite literally universally applicable—holds on Triton and Titan, Neptune and Jupiter, and on other solar systems and faraway exoplanets, as well as locally everywhere (though with no effective wind-related consequences on Mercury and the moon), and stands foursquare as the fundamental basis of why wind blows, and, by more general extension, why winds blow.

Initially, the more detailed explanation seems simple enough. The heat source that plays out the drama for Earth is, naturally, our sun. The atomic fusion reactions that are occurring within what to astronomers is a fairly normal, unexceptional, yellow dwarf star of the main sequence consume a prodigious amount of the sun's mass—about 4.7 million tons of it every single second. The energy created by these reactions pulses unstoppably outward from the solar surface, a great interstellar firehose of electromagnetic radiation that ranges across the entirety of the spectrum, from radio waves to gamma rays, from infrared to ultraviolet, from visible light to X-rays. This star-born benison gushes outward at the speed of light, washing across and through anything in its path. And at that speed—covering 186,000 miles each second—it takes just a scant eight minutes and twenty seconds for the radiant wave to travel from the seething hellscape of the solar surface to the orbit and then the presence of Earth with its encompassing pale blue blanket of the most remarkable and benign and life-giving atmosphere thus far known. This blanket, made up largely of nitrogen, oxygen, and argon—and with smaller amounts of carbon dioxide and in some places water vapor, and with a confection of gaseous oddities, including neon and krypton and (our slow-growing gift from Earth's one billion robustly digesting cattle) methane gas—keeps us and all our colleague animals and plants and bacteria and algae—for the time being—alive.

This atmosphere* then reacts in a series of ways, some probable

* This spheroid-shaped mixture of gases and vapors that entirely blankets the Earth weighs a little more than 5.5 quadrillion tons, bearing down at sea level with a pressure of around 14.696 pounds on every square inch. The Earth, which weighs some 5.92 million times as much as its atmosphere, has sufficient gravitational pull to keep this tonnage of gas adhering snugly and securely to its surface. It does so in five defined layers. The lower four to twelve miles is the troposphere, within which most surface life lives, where most weather occurs, and where most winds blow, and where temperatures fall at a regular "lapse rate" of some 3.56°F for every thousand feet of ascent. Above the boundary, the tropopause, lies the thirty-odd-mile-thick stratosphere—dry (rising clouds stop sharply here), and in which the lapse rate slows and then reverses so that at the next boundary mark, the stratopause, the ambient temperature can be as high as 5°F, survivable for a human were it not for the near-total lack of breathable air and a pressure so low your blood would boil and your arteries explode. Within the stratosphere is the thin but vitally important ozone layer, which forms for Earth and its inhabitants a protective shield against harmful solar radiation. Beyond the stratopause—above which no jets can fly, only rockets—is the mesosphere, within which is the informally sited Kármán line, at sixty-two miles high, nominally regarded as the boundary of Outer Space. Farther still is the intensely cold mesosphere—at its outer edge as chill as minus 120°F, and yet ironically where most incoming meteors burn up—after which is the thermosphere, home to the ionosphere—vital for reflecting radio waves and causing the spectacular light shows of the two auroras, borealis and australis, and home also to the gravitation-free International Space Station—and finally comes the exosphere, where there are countless satellites in orbit but no weather, and absolutely no wind.

and predictable, others more eccentric, with the thunderous arrival of the inrushing radiation from the sun.

The simplest and most predictable reaction has to do with heat, and with the first two laws of thermodynamics. The thermal energy from the sun is converted (without loss or enhancement, as per the First Law) into the kinetic energy of the molecules of all the nitrogen and oxygen and krypton and methane that it encounters. Since the Second Law states that thermal energy must invariably pass from bodies that are hotter to those that are cooler, it follows that the atmosphere encountered by the solar radiation will heat up. Its molecules will begin to move faster and faster; they will separate one from the other as they do so, such that fewer of them will occupy those cubic centimeters of space that once they occupied en masse; and the gas will become not just warmer but lighter too, and so will, according to the inalienable rules of gravity, rise upward. In summary: the solar radiation will warm up the air, and as the air rises it leaves behind what can best be described as an absence, a void, an area of lower pressure than existed when the predecessor air, now wafting skyward, was resting there calmly and comfortably.

And so, nature abhorring a vacuum—a fact first mentioned by Aristotle almost three thousand years ago in his classic *Meteorologica*—air from the cool surroundings of the absence rushes in to close the gap, to heal the wound, to fill the hole, to bring the pressure back into equilibrium and allow the absence to revert once more to a presence. Vertical outflow is replaced by horizontal inflow. And it probably needs not to be stated yet again that it is this inflow of air that is, in its most basic form, the wind.

In the simplest imaginable theoretical arrangement—with a single source of heat being brought to bear on a single column of air—the wind generation would inevitably be just a temporary affair. The warm air would rise, cooler air—as wind—would fill the gap where it had once been, and the warmed air would cool and sink back down again, on and on until all the air in the system had reached the same temperature and all motion came to a stop. In the real world, however, matters are a great deal more complicated, ensuring that the wind keeps on blowing, and in a near-infinite variety of ways.

Not the least because the atmosphere that is being confronted by the onrushing swell of solar radiation is, in common with the spheroidal planet it blankets, also spheroidal in shape. Moreover, it is tilted over to one side at a slowly changing angle. It is in a permanent state of daily self-rotation around this selfsame tilted axis. And once a year it also circles, with the planet it blankets, around the very sun—it orbits it, along an eccentric elliptical path—from which the radiation sweeps and performs its magic. The interplay of these four simple factors—and there are many more that are less simple, and one in particular is crucial to the story of atmospheric circulation—ensures that the winds that blow across the surface of the Earth itself are in a constant state of change, endlessly variable, and showing no signs—no signs *yet*—of wanting to slow or stop. Earth's winds are capricious in the extreme, and all because Earth itself behaves in a variety of capricious and mesmerizing patterns.

Difference is the key. The four cardinal wind-related features of Earth—its shape, its tilt, its rotation, and its orbit—ensure that the sun's radiation falls upon its surface at different angles and with different intensity at different times of each day and each year. At Timbuktu, in the west-central Sahara and close to Earth's equator, the rays fall for the most part directly perpendicular to the surface, making that surface unbearably and unremittingly hot and keeping the air above in a permanent rising mode, as if it were belching up from a factory chimney. In Aberdeen in northeastern Scotland, by contrast, the sun lies much lower on the southern horizon, its rays hitting the Grampian countryside with only a glancing blow, the amount of rising air generally negligible other than at those midday cloudless moments when an unobscured sun manages to produce some local warming. And then thirty-some latitude degrees farther on, at the North Pole, the angle of the sun can in the winter season be so low as to create near-permanent darkness, and the local air is dense and heavy and resists all attempts to rise, except when it passes over those parts of the Arctic Ocean that are a trifle warmer than the rest.

At this point a broad distinction needs to be made, and it happens by chance that the mention of Timbuktu, Aberdeen, and the North Pole allows this distinction to be explained more easily.

There are two broad categories into which all matters meteorological—particularly matters relating to wind—are classified. There are ultra-large-scale events and trends, which are lumped together under the term *synoptic*; and there are much more local occurrences that take place on what is referred to as the *mesoscale*. The physical principle underlying both is just the same—vertically rising warmed air is replaced by horizontally flowing cooler wind. The scale is what distinguishes them. In terms of wind quite specifically, mesoscale is what most concerns us as individuals. Will tomorrow in Aberdeen bring humid air from the south or chilly polar air from the more northerly waters around the Shetlands and the Faroe Islands? Will that line of thunderstorms approaching from the Cairngorms to our west bring any dangers in its train? Will my daffodils be flattened by tomorrow's gales? Why was this just a breeze and that next door a gust? And there are their even more local kin, the matters of the microscale—with the term MMW, signifying mesoscale-microscale weather studies, being a thing these days, a weather world in which puffs of cloud or breezy wafts are of more interest than storm fronts or weeklong droughts. In the meantime, though, the broader question of synoptic wind patterns—the big boys of the planet, which allow us to look at grand-scale matters involving the atmosphere's circulation around the planet as a whole—needs a more detailed explanation. It is between places like the equator and the North Pole that such distinctions become most readily apparent.

The temperature in Timbuktu is invariably hot—it is 105°F at the time I am writing this—and almost precisely freezing—32°F—when I called a few moments later to get the temperature at the North Pole. Five thousand miles separate the two places. What happens to the air that is heated over the blisteringly hot Sahara is in one sense just as expected. It expands, it rises, and then cools as it reaches higher and higher toward the tropopause, and if it had collected any moisture on the way up, it sheds it in the form of rain, then heats up again and rises once more to form great thunderheads as it slowly shifts through the tropics until finally, reaching the tropopause—the boundary with the stratosphere—at around sixty thousand feet, it finds it can go no farther other than to head toward where the air below is assuredly cooler, in the direction of the North Pole. A straight five-thousand-mile shot—a river of pure high-altitude wind hurtling through inner space from the equator to the Arctic; what could be more simple?

More simple, and quite as beautiful. At least in theory. Recalling the fundamental laws holding that energy can be neither created nor destroyed, it is worth considering what the Earth does with all the thermal energy it receives from the ceaselessly radiating sun. We know all too well from history, anecdote, official records, and personal experience that the energy falling onto Earth dissipates itself somehow; if it did not, our planet would have become unbearably hot very many years ago, and life would quite probably never have evolved on either its land surface or beneath its ocean waves. But in fact a contented thermal balance has somehow been achieved, keeping Earth habitable and its habitants happy. And how that balance has been managed is quite simply by virtue of the planet radiating its excess heat back into space—most of it, of course, radiating from the hottest places, like the tropics, but a not insubstantial amount presses outward from where it is coldest, from the poles. And though it might seem counterintuitive to suppose that there *is* any surplus heat available to be radiated back into space at the poles, consider this: although the actual perceived difference in temperature between the High Arctic and the Sahara desert might seem considerable—today some 73°F separates Timbuktu from the Pole as I write these words—in real and absolute terms, the difference is not that great at all.

And that is because the temperature scale used here—Fahrenheit; and for this argument it might just as well be Celsius, or Centigrade, or that vanishingly rare system known as Réaumur—is quite artificial and is based on the change in phases of water, from solid to liquid to gas. That is all very well for humans and our water-based society; but in absolute terms there has to be a scale that originates at a point of the total cessation of molecular movement generally, whether the molecules involve hydrogen or oxygen or iron or gold or uranium—and that is the still practically unattainable point known as absolute zero. A scale that starts here and has no upper limit has existed since 1848, when Lord Kelvin declared that absolute zero was by his calculation 273°C below the freezing point of water, and would henceforward be designated Zero Degrees Kelvin. His eponymous scale would then creep upward at one Kelvin degree for every one Celsius degree. So, 0°Celsius, when ice melts or water freezes, would be redesignated as 273°K; water would boil at a hundred degrees higher, at 373°K; and the temperature in Timbuktu—recorded on this writing day as 105°F or 41°C—would by his Lordship's absolute scale be 314°K.

And herein lies a point that is quite invaluable in understanding the theoretical generation of wind: that in absolute terms, the difference in temperature between the poles and the equator is not that great—though it may seem dramatically different; ice floes and penguins are very different from sand dunes and camels—but the temperature difference in absolute terms, related to the fundamentals of what makes the universe run, is between the 314°K of the Sahara and the 270-odd degrees Kelvin at the North Pole.

At best, this difference is no more than a modest 15 percent. And in terms of radiating energy and dissipating it from the Earth's surface in a manner that will keep the planet from overheating, the poles are quite as radiative as are the equatorial regions. A little bit cooler and so a little less radiative, true—15 percent less. But the old idea that it is far too cold at the poles for any outward radiation to occur at all is quite untrue. The Earth is losing energy—indeed, has to lose energy—at all available points on its surface.

And if this is true, that the poles radiate nearly as energetically as does the equatorial zone, how does the energy get from one place to the other? Tim Woollings, an Oxford meteorologist, summarizes it thus:

> The answer lies in the atmosphere and also, to a lesser extent, in the oceans. This is the most fundamental reason for all the winds and ocean currents on Earth. The atmosphere and oceans have a mission: to relentlessly transport energy poleward, in order to balance the books and to provide energy to the polar regions which can then be lost to space. Most often the mission is accomplished by a simple flow of heat, with warmer masses of air or water moving poleward and colder masses moving equator-ward. Without the transport of heat by these circulations, Earth would not even be able to support life as we know it, as the tropics would be unbearably hot and the higher latitudes impossibly cold.

And yet. Were the planet beneath the airstream quite static, this air would indeed flow without interruption or complication. But the planet is not static, as we know. It is instead spinning, taking one full day to complete one full rotation—the very definition of a day, in fact. Viewed from above its axis, the Earth is spinning in a counterclockwise direction, and so, since it is a sphere (or spheroid, oblate in shape), it does so at different speeds at different places on the surface. To someone standing at the twenty-five-thousand-mile-long equator, it will be spinning eastward quite fast, at a little over a thousand miles an hour. A Malian in Timbuktu, at 17 degrees north, will be going at a little less than nine hundred miles an hour. A Scot standing on a headland in Aberdeen, which is nearly 60 degrees north, will cover thirteen thousand miles in each rotation, going at 550 mph; and someone standing at either of the poles will not cover any distance at all but will merely rotate once daily, effectively going nowhere. This difference in the speed of different parts of the planet—faster at the equator, slower as one goes farther north or south—produces a most interesting set of phenomena.

It would play havoc, for instance, with that just-mentioned and fancifully conceived river of warm equatorial wind that is moving poleward from the tropics, such that were this wind to put itself down anywhere along its route—a route it believes to be a straight line—it would find that it was in fact many miles to the east of where it had expected to be. To a neutral observer it had moved not in a straight line at all but along a curve to the right, a curve that had become progressively more curved the farther along its path it had progressed. A great and disconcerting complication had arisen and apparently quite altered the airflow's direction. This is famously known as the Coriolis effect.

AROUND THE FOUR sides at the top of the massive first stage of the Eiffel Tower are inscribed in two-foot-tall gold lettering the names of seventy-two French scientists and engineers whom Gustave Eiffel considered preeminent and worthy of perpetual memorial. Among them are a good number of familiar names: Daguerre (of photography), Fourier (mathematics), Fresnel (lighthouse lenses), Coulomb (electromagnetism), Becquerel (radioactivity), Cuvier (zoology), and, listed on the tower's southwest side, close by the name of a much-forgotten man who invented a means of digging through waterlogged earth and another who perfected a kind of dirigible airship, is that of Gaspard-Gustave de Coriolis, a slender, clean-shaven and short-lived Parisian physicist who in 1835, eight years before his death at fifty-one, came up with an explanation that quite revolutionized our understanding of the sometimes curiously inexplicable behavior of the Earth's atmosphere.

The paper in which he explained the mathematics behind the pair of eponymous physical phenomena that are his indelible legacy—the Coriolis force and its consequence, the Coriolis effect—was actually concerned not with weather at all but with the mechanics of waterwheels. By the time the meteorological community got wise to the relevance of his writings, Coriolis was long dead. But his explanation has proved enduringly satisfactory, displaying both why prevailing winds blow as they do—as with the westerlies that waft over the Atlantic Ocean—and why Northern Hemisphere cyclones rotate as they do, counterclockwise, while those south of the equator rotate in the opposite direction.

Put as simply as possible, and considering that planet Earth, where this is all happening, is rotating in its customary counterclockwise direction, consider what will happen to an object that is thrown in a due northerly direction, toward Aberdeen, from a point due south of Aberdeen on the equator. At the originating point—which would be somewhere at sea, south of Nigeria—the object, as well as the planetary sea surface underneath, will be moving, as explained, always toward the east, at rather more than 1,000 mph. Up in Aberdeen, the surface will also be moving eastward, but much more slowly, at around 550 mph. The Coriolis force will in consequence act upon the thrown object and nudge it slowly and steadily to the right, to the east. It will never land in Aberdeen. It might well end up in Copenhagen, or even Moscow, the scale and degree of its deflection being the result of the application of the force, and thus the Coriolis effect.

Were this hurled object not a simple physical entity—a ball, or a shot from a cannon, or a missile, or a magical flying paper dart, all things that are imagined in various depictions of matters relating to Coriolis—but a warm wind trying to make its way northward from the equator in the direction of the North Pole, then the Coriolis force would trigger the Coriolis effect as a giant aeolian intervention. The northward streaming wind would apparently turn and turn and turn according to its inertial nudging until, some many degrees of latitude north—somewhere off the coast of Morocco—it would be blowing no longer in a northerly direction but toward the east. It would have transmuted itself, in the irritatingly illogical custom of the meteorological priesthood, from a southerly wind into that most famed of transatlantic winds, a westerly.

Would that it were that simple. And before launching into an explanation of why matters are a little more complicated, it is perhaps worth a reminder that we are dealing here with big wind patterns, with synoptic winds, with matters relating to global atmospheric circulation. These are winds that affect the world in its entirety, and that in human terms have and long have had implications for the development of human history. Questions relating to such phenomena as weather fronts and squalls and gusts and on- and offshore breezes and williwaws and foehns and katabatic winds and thunderstorms and cyclones—in other words, mesoscale events, as already noted—are only peripherally related to the complicated nature of the general atmospheric circulation. These factors involve two men whose names are forever enshrined in meteorologic lore: an eighteenth-century Briton named George Hadley, and a nineteenth-century American named William Ferrel.

Hadley was trained as a lawyer but was by inclination a meteorologist, intrigued in particular by data that was by then—post-Enlightenment, when scientific observation had become very much the fashion—pouring into the newly founded scientific establishments in Northern Europe, especially in London and Scandinavia. He joined the Royal Society and was given the task of crunching the weather-observation numbers that were coming in abundance from ships and weather stations—and from the back gardens of amateurs who, like himself, found weather a most fascinating topic.

What intrigued Hadley most were the logs from ship masters who had recorded the winds in the central Atlantic Ocean—between 30 degrees north and 30 degrees south—and that (except for the windless purgatory of the doldrums, *qv*) seemed to be blowing steadily, relentlessly, no matter the season, in one prevailing direction—toward what one might loosely call the west. More specifically, those winds blowing north of the equator came from the northeast, while those south of the line blew from the southeast. Ships' masters called these regular, unfailing, and satisfyingly brisk breezes *trade winds*, since they were commercially highly advantageous to any vessel wishing to get as quickly and expeditiously as possible across the sea, from Lisbon to Barbados, say, or from Senegal to Recife. And that was just across the Atlantic: logs from vessels that ventured even farther afield, into the immense wastes of the Pacific Ocean, showed that the selfsame belts of winds blew there as well, and just as constantly. Hadley was able to determine from all those submitted logbooks too that northeast trades and southeast trades blew between California and Borneo, between Cartagena and Batavia, and happily took great sailing ships—most of them in the early eighteenth century kept busy exploring and conquering*—along with them as well.

As to why such winds blew, many were puzzled. George Hadley offered the most plausible explanation. Of the many who had been perplexed, the best-known was perhaps the great astron-

* In the Atlantic Ocean, much early use of the northeasterlies was made by those running the unsavory business of the "triangle trade"—slaves brought from Africa to the Americas, cotton and rum hauled from the Americas to Britain, metals and trade goods from British ports to west Africa. The wretched westbound cargoes were by far the swiftest; the Britain-bound ships headed north to catch the prevailing westerlies above the 45th parallel—proving, though not realized at the time, the exact dimensions of what, as a later memorial to George Hadley's breakthrough theory, would be called the Hadley cell.

omer (and Astronomer Royal) Edmond Halley, for whom the famous comet is named. His idea, which briefly caught the public attention, held that the sun, heading ever-westward, quite literally dragged the winds behind it, employing some unspecified quasi-magnetic attraction on the atmosphere. Hadley, who as a lawyer did not wish to seem impertinent to Halley, who at the time had the status of a national treasure, nonetheless offered up, with carefully modulated reverence, the notion that solar dragging was nonsense. Instead, Hadley explained in a 1735 paper of a modest four and a half pages (in what would in time become the world's longest-running academic journal, the *Philosophical Transactions of the Royal Society*) just what was happening. The equatorial air broiled by the sun would rise high into the sky, perhaps as much as six miles high, and, in that part of the world lying to the north of the equator, it would head northward toward the cooler air of the North Pole; while to the south of the same line, a great river of similarly high-flying hot equatorial air would head south toward Antarctica.

At some point along its odyssey the hot air would start to falter and lose altitude until it dropped down toward sea level and joined the river of cooler near-surface air that was already, much farther south, rushing to fill the void left by the rising hot air. And what George Hadley noted was that this return stream of cooler air, running so close to the earth's surface—and, crucially for the world's mariners, running so close to the *ocean* surface—would for some reason run in a *southwesterly* direction north of the equator, and *northwesterly* on the other side of the line. Something connected to the earth's counterclockwise rotation nudged the wind off course. What might in ordinary circumstances be a simple southerly airstream would have been nudged and nudged the farther it progressed.

And eventually these became the northeasterly and southeasterly trades—constant, regular, powerful, of great utility to the growing armadas of seaborne world trade. Trade's original long-distance land-based mechanisms—the camel trains, most notably—cared little for which direction the wind blew en route. To a caravel, a carrack, or a clipper, wind was everything.

As to why the trades blew as they did, in the directions they did, Hadley could not rightly say. He might have wished to know of the intricacies of linear and angular momentum, but it would be a further century, almost exactly—in 1835—before Coriolis adduced the principles of the force that bears his name. Hadley knew nothing of these mysteries and was flying blind when he wrote his 1735 paper. But he got it right—which is why his name is memorialized still, and will be forever linked with trade winds and why they blow as they do. Sir Napier Shaw, one of the high priests of the meteorological world,* said of George Hadley, perhaps a little clumsily: "The theory of the trade winds which Halley started and Hadley improved belong to the fairy tales of science, because they explain the complexity of nature by a simplicity which is suggestive of a fairy's wand."

If Hadley is forever linked to the trades, then William Ferrel, a barely educated farm boy from rural West Virginia, owns the explanation of that other great component family of the world's atmospheric circulation, the westerlies. We were taught endlessly at my boarding school that Britain's weather was gently moderated by the five Ws, the "Warm Wet Westerly Winds in Winter." It would fall to young William Ferrel to explain why such westerly surface winds blew as they did—an achievement that explains why he and George Hadley are eternally connected, their names bestowed upon a pair of critically significant atmospheric features, the Hadley cell and the Ferrel cell.

Ferrel's explanations were almost wholly based on mathematics—not for him the examination of ships' logs that had so intrigued

---

* He was a particular expert on smog, and wrote a definitive book on the problem of smoke in cities.

Hadley. His particular interest was in the origins of the westerly surface winds. He knew why the trade winds blew—but why the westerlies, which seemed at first blush to have little to do with the great uprising of heated tropic air that appeared to set the whole mechanism in motion. His epiphany came in 1856—the weather community had been wrestling with the puzzling idea for well over a century—with a paper that he published in, of all things, a medical journal in Nashville, Tennessee.

His idea stated here demolished, quite literally, the idea that the global circulation pattern was based on the notion of one gigantic heat-driven atmospheric escalator, with one side rising up and outward and the other falling and returning home in an endless world-girdling churn. Instead, Ferrel declared to Tennessee and far beyond, there are in truth *three* distinct and self-contained areas of convective churn. Three cells in each hemisphere, functioning in concert to dictate the basic patterns of the atmosphere's shiftings, determining the world's fundamental weather.

Down by the equator, Ferrel calculated, upwellings would force the warm air through some 35 degrees of latitude before it would sink down and return at the surface as trade winds. Mathematics then suggested there also be, similarly, smaller convective cells headquartered at the two poles—with warmish air heading toward the zero latitude mark, then cooling and sinking and building up the pressure around the poles themselves, from which would then pour cold blasts of east winds (east because of the Coriolis force), which would make everyone suffer through them—think Napoleon's forces retreating from Moscow, or the misery of the Allied wartime convoys battling against the icy northeast blasts while heading from Iceland to such Soviet ports as Murmansk and Archangel.

And between these two sets of cells, those by the poles and those by the equator, would be the third, which mathematics said would operate—and observation later confirmed did operate—in the temperate zones between latitudes 35 and 60 in each of the hemispheres. Air would rise and fall in a convectional manner, as expected—but here the warmer winds would not push up to high altitude on any long journey toward the poles but instead make a much shorter trip while remaining at the surface. The high altitude return winds in these zones seemed to be of much less consequence, since humanity is only directly affected by winds that blow at the surface. And so here in the temperate zones it is westerlies that dominate everything, especially in the Northern Hemisphere, where landmasses—America and Europe, most notably—stand directly in their path.

The three cells are now formally accepted as dominating the world's circulation pattern. Since 1955 the meteorological community has known them as the polar cell, the Ferrel cell, and the Hadley cell. On further analysis each cell offers up the answers to a variety of smaller-scale phenomena in great detail, but their offering of neat and tidy reasonings behind the existence of the world's two best-known winds, the westerlies and the trades, is still regarded as most satisfying of all.

Before leaving the atmospheric pattern language on this synoptic scale, one further part of the planet's grand wind design remains to be explained. This was a phenomenon not known to exist until the late 1940s. It was, however, suspected some while before. It was a mysterious wind that seemed to operate at very high altitudes and had little to do with the equatorial and polar upwellings and surface winds that would so intrigue the nineteenth- and early-twentieth-century weathermen who were trying to work out the mechanics and geography of the atmosphere's circulation. One notable Midwestern astronomer-mathematician, Elias Loomis (whose tiny observatory, one of America's oldest, still stands in the town of Hudson, Ohio), came up with the idea that some powerful force, possibly unobservable from the Earth's surface, was responsible for driving the summer storms across the American plains: those that caused the debilitating agricultural pain now remembered as the Dust Bowl, windstorm after windstorm in ceaseless trains of misery—what caused them to move so, Loomis asked; maybe something unseen, high above the normal wind patterns experienced down by the surface.

Then came the fantastically powerful eruption of the island volcano of Krakatoa, in the Dutch East Indies, in late August 1883, which gave a boost to the gathering notion that something inexplicable was going on at ultra-high altitudes. Microscopic silicate particles from this immense blast were thrown high up into the troposphere and may well have penetrated the tropopause and into the lower reaches of the stratosphere, where they were promptly swept, at exceptional and unanticipated speed, right around the world. The effects were dazzling—literally, with vividly colored skies and truly wondrous sunsets delighting, and sometimes alarming, people who knew nothing of the faraway eruption, nor could imagine any explanation for the phenomenon. Some have claimed, as already mentioned, that the chromatically anarchic skies that form the background in Edvard Munch's famous painting *The Scream*, composed a decade later, were prompted by his memory of Krakatoa's dust over the Norwegian fjord country.

A more logical and observation-based explanation for the mystery would come nearly half a century later from a Japanese scientist named Wasaburo Ooishi, who worked at the Japanese Aerological Observatory on the flanks of the famous twin-peaked Mt. Tsukuba, forty miles northeast of Tokyo. From here he launched a number of radiosonde balloons with the aim of recording the behavior of the atmosphere at ever-increasing heights. In almost all cases the balloons and the transmitting instrument packages suspended beneath them encountered a sudden torrent of very fierce westerly winds, which blew the balloons eastward and out into the Pacific Ocean, where they were lost. The winds blew consistently, Dr. Ooishi discovered, once the balloons had ascended through some sixty thousand feet. The evidence was incontrovertible: some force, unknown among the surface winds, was provably operating ten miles above the earth, and he, Dr. Ooishi, had discovered it. He would write up his results and observations immediately, to take—as academics are quite reasonably wont to do—historic credit for it. Accordingly, he sat down and wrote—but neither in Japanese nor in English. He wrote in Esperanto, a language invented by a Russian ophthalmologist in what at the time was Poland. The paper, written in this then generally unfamiliar (though still surviving) tongue, was headed *Raporto de Aerologia Observatorio de Tateno*, and though modestly triumphal in tone and, in the history of world meteorology, seminal, it went almost entirely unread. Few explanations have surfaced for Dr. Ooishi's dreamily idealistic decision to write in Esperanto, other than his pride in having recently been elected board president of the Japanese Esperanto Institute. This august body, like the language it seeks to promote, still exists today and presides over eighty local chapters throughout Japan, and it publishes *La Revuo Orienta* to help keep local enthusiasm high.

Undeterred by the deafening silence that greeted his announcement, and unassailably dedicated to his linguistic evangelism, Ooishi wrote a further eighteen papers on the same topic, all still in Esperanto, before dying in 1950. He was honored by his country with the Japanese Order of the Sacred Treasure, but his standing in the world of wind is little more than a place in the footnote jungle of learned books about the world's weather.

Ironically, the generally accepted modern discovery of the wind that would soon come to be known as the *jet stream* came about in Japan also, but under dramatically different circumstances.* It was in November 1944, at the beginning of the Amer-

* Late in the war Japanese scientists came up with the elegant idea of using Ooishi's jet stream to speed bomb-carrying balloons eastward across the Pacific Ocean toward random and unspecified targets in the western US. They called the weapons *Fu-Gos*, or "balloon bombs." Starting in November 1944, some 9,300 such weapons—mulberry paper and persimmon-paste balloons thirty-three feet in diameter and carrying one anti-personnel bomb and a variety of incendiary

ican bombing raids on mainland Japan that, by today's general admission, were cruel and unnecessary and probably classifiable as war crimes, and which sought to devastate an already impoverished, demoralized, and near-defeated nation and were intended, as US general Curtis LeMay was later to say about the North Vietnamese, to bomb Japan "back into the Stone Age."

The Army Air Forces plan was simple enough. By this time in the war American forces had succeeded in gaining access to the Mariana Islands, and Navy Seabee construction battalions had quickly thrown up a rudimentary airfield on the southern tip of the Marianan island of Saipan. From here it was only 1,500 miles to Tokyo, a manageable distance. The Army Air Forces now had access to near limitless numbers of newly built Boeing B-29 Superfortress heavy bombers, and with their near-four-thousand-mile range, fully loaded with bombs, it should in theory be possible for squadrons of the aircraft to rain hundreds of tons of explosive ordnance down on strategic Japanese targets and return safely to base, having day by day and night by night wrought terrible damage, factory by factory, on the enemy's war-making abilities.

The first major operation, conducted by the 73rd Bombardment Wing of the Twenty-First Air Force, took place on November 24, 1944. It proved an embarrassing shambles. One hundred eleven Superfortresses took part, flying at around 230 mph in a northwesterly direction, making a straight-line passage from Saipan to Mount Fuji, some fifty miles to the west of the Japanese capital. Here the squadron would regroup and turn due east

---

devices to set American forests ablaze—were launched into Hokkaido's upper troposphere. Though one reached as far as Michigan, most fell harmlessly into the ocean. However, on May 5, 1945, a group of five Sunday school children led by one Elsie Mitchell, picnicking in the forest near Bly, Oregon, came across one of the grounded objects and a child reached to turn it over. It exploded, killing all six. The Mitchell Monument and the nearby shrapnel tree, peppered with steel fragments, commemorate the tragedy.

toward the city of Musashino, a few miles northwest of Tokyo, where the Nakajima Air Engine Factory was the Pentagon's designated target.

Things started to fall apart as soon as they arrived at Mount Fuji. Seventeen of the planes had had to turn back to Saipan, most of them with engine issues. Six others then found themselves unable to drop their bombs. There were dense clouds all across the capital, extending to Musashino. And, most inconvenient of all, these high-flying Superfortresses found themselves, when they reached their bomb-run altitude of thirty-six thousand feet, in the iron grip of a wholly unexpected gale-force westerly wind that propelled them over their factory target at more than 450 mph—much, much too fast for the bombardiers to get a fix on the factory below.

Only 24 of the 110 bombers managed to attack the Nakajima factory, and only 48 of the 240 bombs they carried landed anywhere near their intended targets—with a dismaying number of the bombs turning out to be duds.

The planes returned to Saipan, and while recriminations were plenty and bombing plans changed as a result, the consensus was that the high-level westerly winds, unpredicted by the weather forecasters, were the principal culprit. It was readily acknowledged on Saipan, at Pacific Theater HQ back in Hawaii, and in Washington, DC, that what the B-29s had experienced was a brand-new wind, on which research could and must commence immediately.

The jet stream had officially been born and universally acknowledged—and this time, some eighteen years later, in English. As its existence was confirmed and mapped and analyzed, the realization dawned that it played a crucial role in the direction of the world's weather—quite literally: where the jet stream prods it, so goes the weather below. It doesn't cause the weather; the making of it belongs in the realm of the mesoscale events mentioned occasionally so far. But in determining where a whole host of meteorological dramas are persuaded to go, the jet stream is the villain.

"It," however, is not the proper pronoun. "They" is the more appropriate, since it was quickly discovered that there were in fact four jet streams, all similar in aspect, being narrow and extremely fast and very cold, and all blowing from west to east. They were found to be spread across the globe as matching pairs, north and south of the equator. Indeed, once the approximate positions of the three great convection cells—the Hadley, the Ferrel, and the polar—had been established, it was noted that the jet streams occurred, or were based or mainly operated along, the two lines where these three cells met.

The upwelling warm equatorial winds of the Hadley cell begin to cool and dip downward at around 30 degrees north latitude—which is where the cooler reverse winds, the trades, start their commercially valuable westbound journeys. At this very same point, 30 degrees north latitude, the warmish surface airs of the Ferrel cell are heading west across the oceans, arriving eventually at around 60 degrees north, where they hit the downslope winds from the polar cell and turn to begin their own upper-atmosphere odyssey back across the intervening 30 latitude degrees to where they started, at the edge of the Hadley cell. It is at these two junction points—the Hadley-Ferrel cell junction, and again at the Ferrel-polar cell junction—that the two great jet streams begin to slither and rage around the planet. Two in the Northern Hemisphere, and two farther to the south of the equator, each of them heading relentlessly eastward, since, as is axiomatic, the jet streams are eastbound winds, their points of origin still the subject of endless mathematical theorizing and calculation. As to exactly why they form, the jury remains sequestered. As to what they do, that is now quite well known—and, in a single phrase, can be summed up as *they wreak havoc.*

There is one place, one region, where a jet stream can actually be seen: the Nepal Himalayas. Tectonic forces have ensured that a scattered few of the great eight-thousand-meter peaks there—Lhotse, Makalu, K2, Kanchenjunga, and, most symbolically of all, Everest—thrust their limestone summits up into the very lower reaches of that layer of the troposphere where the equatorial jet stream operates. When it dips and curtseys low enough, its effects on the peaks are visible, particularly on Everest's eastern slopes—a thin bridal veil of blowing snow can be seen drifting across the vivid blue sky toward Lhotse or to Makalu, a vision at once serene and seemingly fragile.

Up where it blows, the reality is anything but. The wind speed, bitter cold and howling with menace, reaches into the scores of miles per hour, occasionally topping a hundred, and the long lines of climbers trying for the summit as it blows can be easily discouraged by their brush with it. In recent years scientific expeditions have constructed a small number of automated weather stations in these inhospitable parts, in a bid to keep track of the gales and to find out more about these enigmatic high-speed snakes of air. One

The snow blown eastward off the summit of Mount Everest is being transported by the jet stream, in one of the few places on the Earth's surface directly affected by this normally high-altitude air current.

of the stations, heroically and painstakingly assembled by teams of sherpas on an ice ledge and called the Beacon, was tumbled over, wrecked, and buried by just a year's worth of accumulated ice and furious westerly blasts. A salvage party went back up a short while later and bolted firmly to a foundation what all hope will prove an even more durable station sited on a rock platform just a hundred feet below Everest's 29,029-foot summit. At the time of writing it appears to be in good order, sending jet stream data via satellite from what was thought to be the world's only fixed observatory capable of measuring its behavior.

However, on the Tibetan side of the frontier that technically runs across the summit of Everest, the Chinese government has been building a chain of weather stations on the slopes of the peak they called Qomolongma. Seven have gone up so far. With the near-exponential growth of Chinese science it is assumed that Beijing will soon know more about the mathematics and future behavior of the jet stream than meteorologists elsewhere.

Airliners feel the influence of the jet stream, invisible though it might be to the pilots who navigate across and along it. On the busiest east-west routes—across the United States, across the Atlantic, the overnight runs from Europe to the East Asian capitals—the eastbound flights riding the jet stream are invariably much faster than those heading against the streams' prevailing direction. In February 2024 the National Weather Service in Baltimore announced that its radiosondes were recording jet stream speeds of more than 250 mph at altitudes above thirty thousand feet—and as a direct result of riding the stream, a flight from New York eastbound to London and another from Newark to Lisbon clocked over-the-ground speeds well in excess of 750 mph, ostensibly moving faster than the speed of sound.*

* Not breaking the sound barrier, though, since the planes were embedded in an envelope of fast-moving jet stream air, relative to which they were merely lumbering

And not infrequently, unusual and unpredictable jet stream activity can have a direct effect on the surface weather miles below.

When in the late 1970s I lived with my family in India, we liked to take holidays next door in Himalayan Pakistan, specifically in the Swat valley in the far northwest of the country. We would drive there in the family Volvo station wagon, which had already proved its amiable durability a year before when I drove it to India from London, back when it was quite easy to do so. Indian customs regulations at the time obliged me to drive the car outside the country's borders every six months to avoid having to pay formidable extra duties, and so I would make the easy drive from Delhi, where we lived, up the Grand Trunk Road to Amritsar, cross the border at Wagah, head on to Lahore and thence up the GTR to Attock Fort, and after turning right and heading up into the hills along the Swat River via the Malakand Pass, to the regional capital, Saidu Sharif. Beyond, once the hills became mountains proper and there were snows on the high peaks and edging the ancient stone walls that defined the golden water meadows, there can be few more lovely places in creation.

But the idyll was to be rudely interrupted in July 2010. The weather turned dangerously wild. Sudden torrential downpours and an unusually energetic melting of the Himalayan snowpack caused flooding along the Swat River on a titanic scale. Whole villages were swallowed up by the onrushing torrents, farms were entirely inundated, suspension bridges ripped away, the thousands of peach and apple orchards that had brought a measure of prosperity to this most remote of regions were destroyed, the trees drowned in an ocean of glacier-born mud. Hundreds were killed, thousands made homeless. Pakistan had not once been visited by

---

along at their normal cruising speed of around 550 mph, well below the notional Mach 1 threshold.

such rains—not even during this, the height of the monsoon season farther south, during which such rains as fell on the Himalayan foothills were slight, expected, and welcome. If this rainfall was a part of the monsoon, then it was unprecedented, since no monsoon of this strength had ever percolated this far north. It had not done so since the country's founding in 1947; and the weather records of the British India that preceded it showed that nothing remotely similar had occurred for decades before.

So what had caused this disobliging situation in northern Pakistan? A clue as to the ultimate cause lay some two thousand miles away to the northwest of Swat, in, of all places, Russia's capital city of Moscow. For while Pakistan's tragedy that July was the deluge of an extraordinary amount of monsoon rain, Moscow's trials were occasioned by an equally ferocious amount of scorching sunshine—a heat wave that brought the city to its knees, in a tramline-buckling, asphalt-melting, perspirationally challenging two weeks of misery. Much of Central Europe was trapped under the heat dome that developed and then stubbornly persisted, so that forest fires and deaths from heat exhaustion and something like a thousand drownings of people who, frantic to stay cool, had jumped incautiously into water that was cold enough and deep enough to outrank their abilities to swim.

The weather maps then told a simple-enough story. Moscow and its environs were trapped under a huge high-pressure mass of anticyclonic air, circulating hotly and leisurely in a clockwise direction. Swat and its normally more idyllic neighborhoods were, by contrast, trapped in an enormous and somewhat faster cyclonic swirl of air that was moving, as Northern Hemisphere cyclones are by definition wont to do, in a counterclockwise direction. This storm system warmed the Himalayan snowpack and brought cool, wet air to rage down on the villages below.

Cartographic evidence elegantly displayed that two systems, though two thousand miles apart, were seen, because of their gigantic respective sizes, to be somehow meshing with each other like cogs in some diabolical atmospheric weather machine, their outer bands invisibly stitched together in concert, atmospherically connected reciprocals of one another, the yin and the yang of that corner of the global circulation pattern. And what appeared to be connecting and linking and meshing the two systems was an unusual looping and up-and-down diversion, a north- and south-shifting excursion from its normal glide path of the usually reasonably stable and unidirectional tropical jet stream. (The polar jet stream that flows 30-odd latitude degrees to the north and that, as already noted, adds or subtracts ground speed from transatlantic airliners played no part in this story.)

That slender tube of cold, eastbound high-level air suddenly, in the summer of 2010, dipped and rose precipitately through the latitudes, pulling a high-pressure anticyclonic air mass up from the Mediterranean toward the Moscow plains, and at the very same time pushing a chill cyclonic storm-laden airmass down from its normal home above the Siberian pine barrens, its low pressure luring drenching monsoon air up from the Arabian Sea and causing rains to fall with a vengeance upon the downslopes of the normally placid high hills and valleys of Pakistan.

This sudden change of range of motion, the sort of sinusoidal tremor of whiplash that makes the jet stream—or any of the four jet streams, that is, even if only the northern tropical version was affected here—cause events such as those of July 2010, was locally unexpected, and not forecast. But since 1939 the phenomenon has been definitively recognized—one of a series of occasional sudden changes in behavior, direction, altitude, strength, and speed of a jet stream—as being of great importance in determining the behavior of the surface winds that batter or swelter those living below, who likely are wholly unaware of even the existence of these great rivers of air that run so high above them. The changes in the parameters of the jet streams, the nature and frequency of their wild undulations, introduce us to a figure who is every bit as significant to the story of wind as are George Hadley and William Ferrel. He is an

American mathematician-meteorologist of Swedish origin, Carl-Gustaf Arvid Rossby. And what linked and to a degree caused the Muscovite misery and the Swati floodings of 2010—what caused their synchronicity, at the very least—was, we now know, the formation of a phenomenon that would become a classic example of a Rossby wave.

Such waves are enormous, planet-scale undulations in either the upper atmosphere or the deep ocean waters. When, as seen from above, one or another of the jet streams whiplashes toward the poles, the wave peaks are called ridges; and when they force weather systems to dive toward the equator, they are, not unexpectedly, called troughs. The origins of these very-large-scale movements of air (or, in the oceans, water) have much to do with heating and cooling, of course; with the Earth's rotation, of course; with the seasonal changes in the angle of incidence of the sun's

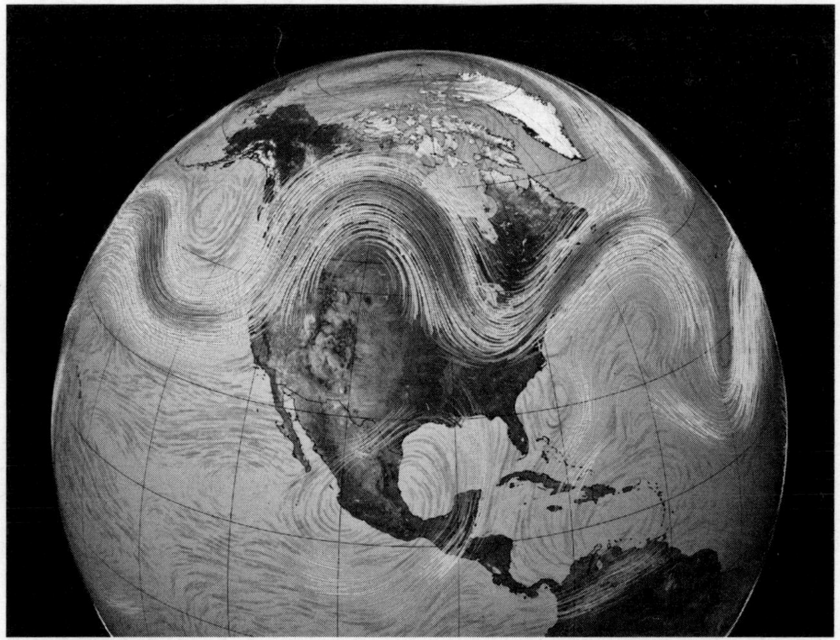

The paths of the various and generally eastbound jet streams, cumulatively assembled from high-orbit satellite data.

rays on various parts of the planet, of course; with the presence of extremely large geographical and topographical boundary features of the planet—the great oceans, of course, and in differing degrees depending on their geographic orientation, the mountain chains. The north-south-running high ranges—the Rockies, the Andes, the Great Dividing Range in Australia, the Urals in Russia, and the Scandinavian ranges in Northern Europe—naturally interrupt and distort and frustrate or cause stress in the supposedly seamless eastward flows of the jet streams. By contrast, the

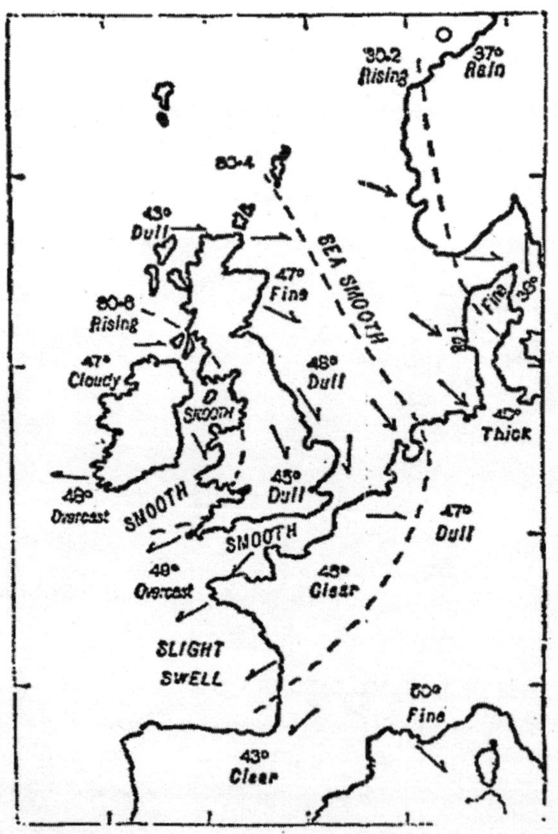

Francis Galton, the Victorian polymath whose reputation has been forever tarnished by his invention of the term eugenics, was an avid meteorologist and drew the first weather forecast maps, published daily in the London Times.

east-west ranges—the Himalayas, the Alps, the Pyrenees, the Carpathians, and, in Antarctica, the Transantarctic Mountains—act to channel or to bounce or to reinforce, like the ice walls of a luge run, the airstreams that pass by overhead.

The introduction of Carl-Gustav Rossby, whose name is now firmly annealed onto these immense undulations, either marks or is coincident with the arrival of the modern approach to meteorology and to its most significant single palpable component, the wind. And of today's approach it can fairly be said that a fantastic and bewildering complexity now abounds.

It was not so back when Aristotle first defined the science of weather and laid out its basic principles three thousand years ago. Nor was it so when, in 1875, the controversial statistician Sir Francis Galton* drew, and had published in the London *Times*, the first-ever publicly accessible weather chart, of the kind familiar today to newspaper readers and television viewers around the world.

But in those early days of meteorology, when so much data was flooding in from balloons and ships and explorers and backyard amateurs who pored over Galton's maps and purchased observing equipment for themselves, and during the years following when theories were being advanced and tested and repudiated or confirmed—in those days it was recognized that to make real

---

* Sir Francis Galton, a towering figure of Victorian intellectual life and still a hero to many, is doomed to be forever tainted by his association with social Darwinism, his belief in the heritability of intelligence, and most notoriously with the pseudoscience—he gave it its name—of eugenics, by way of which much evil was later done, especially in Nazi Germany. But Galton also championed the police use of fingerprinting, studied the efficacy of prayer (finding none), drew maps showing the distribution of beauty around the British Isles (the city of Aberdeen fared worst), invented a whistle that could rate a subject's hearing ability, and devised the most sensible ways both to make tea and to cut a circular cake. His work on popularizing meteorology and forecasting the weather remains perhaps his most enduring legacy.

sense of it all required very much more than the possession of a weathervane, a thermometer, and a barograph—and, in the case of wind specifically, a spit-damped finger raised in the direction of a breeze was needed, as was, in spades, mathematics. In 1922, a British Quaker mathematician named Lewis Fry Richardson famously remarked that so convoluted and multidimensional and shape-shifting had the interpretation and analysis of weather become that "to compute properly usable results by numerical methods would require a year's worth of work of 64,000 mathematicians calculating 24 hours per day." He took this outlandish thought and ran with it, producing an idea of what the sophisticated weather forecasting institute of the future might look like. He named it the Forecast Factory.

> Imagine a large hall like a theatre, except that the circles and galleries go right round through the space usually occupied by the stage. The walls of this chamber are painted to form a map of the globe. The ceiling represents the north polar regions, England is in the gallery, the tropics in the upper circle, Australia on the dress circle and the Antarctic in the pit.
>
> A myriad computers [in this writing, *people who compute*] are at work upon the weather of the part of the map where each sits, but each computer attends only to one equation or part of an equation. The work of each region is coordinated by an official of higher rank. Numerous little "night signs" display the instantaneous values so that neighboring computers can read them. Each number is thus displayed in three adjacent zones so as to maintain communication to the North and South on the map.
>
> From the floor of the pit a tall pillar rises to half the height of the hall. It carries a large pulpit on its top. In this sits the man in charge of the whole theatre; he is surrounded by several assistants and messengers. One of his duties is to maintain a uniform speed of progress in all parts of the globe. In this respect he is like the conductor of an orchestra in which the instruments are slide-rules and calculating

> machines. But instead of waving a baton he turns a beam of rosy light upon any region that is running ahead of the rest, and a beam of blue light upon those who are behindhand.
>
> Four senior clerks in the central pulpit are collecting the future weather as fast as it is being computed, and despatching it by pneumatic carrier to a quiet room. There it will be coded and telephoned to the radio transmitting station. Messengers carry piles of used computing forms down to a storehouse in the cellar.
>
> In a neighbouring building there is a research department, where they invent improvements. But there is much experimenting on a small scale before any change is made in the complex routine of the computing theatre. In a basement an enthusiast is observing eddies in the liquid lining of a huge spinning bowl, but so far the arithmetic proves the better way. In another building are all the usual financial, correspondence and administrative offices. Outside are playing fields, houses, mountains and lakes, for it was thought that those who compute the weather should breathe of it freely.

Gimlet-eyed number-crunchers soon thereafter became the drivers of the new fields of wind and weather science—dreamy pacifists like Richardson or visionary Scandinavian scientists like Rossby were swiftly elevated into the meteorological aristocracy; and for good or ill, weather—for so long so outwardly romantic a calling—has ever since fallen firmly within the penumbra of higher mathematics.

In November 1950 the world's very earliest large mainframe computer, the University of Pennsylvania's ENIAC, the Electronic Numerical Integrator And Computer (it was one hundred feet long, weighed thirty tons, and had 18,000 vacuum tubes and 7,200 crystal diodes) took up the challenge of writing a Richardson-style numerical forecast for the likely wind and weather patterns for each twenty-four hour period during the Thanksgiving week in Princeton, New Jersey. Coincidentally, each forecast took the giant machine twenty-four hours to compute, meaning that it was just, and only just, able to keep up with the very weather it was forecasting. But it was the beginning of a mathematical science that has progressed near-exponentially ever since, its progress largely a consequence of the almighty advances in computing power.

The major weather institutions around the world now all have—and are indeed obliged to have—fully functioning supercomputers to deal with the multitude of data points necessary to offer accurate and long-term forecasts for the weather around the world. No longer does science keep up with the weather: it precedes and predicts it with ever-increasing accuracy.

Only a very few now bemoan the loss of the words and phrases of weather lore—like that of those North Carolina fishermen who still today cleave to the belief that *the third day of a southwest wind will be a gale, and the wind will veer to northwest between 1:00 and 2:00 a.m., in winter, with increasing force.* Or those who still follow Plutarch's remark, made two thousand years ago, that *the Zephyr makes wine ferment more than any other.* Nor Vice Admiral Robert FitzRoy's observation that *strong winds are more uniform and regular than are light breezes.*

Admiral FitzRoy should know. In 1831 he commanded HMS *Beagle* on her five-year, forty-thousand-mile circumnavigation of the world, the voyage and its immense series of scientific discoveries and achievements made famous by its celebrated principal passenger, Charles Darwin. FitzRoy was a pioneering meteorologist of considerable note, founder of the British Meteorological Office, an expert on gales (he insisted on posting gale warnings at British fishing ports, preventing fleets from leaving until the weather calmed, and credited thereby for saving the lives of hundreds), and the man who is reputed to have coined the phrase *weather forecast.*

New Zealand—over which Admiral FitzRoy was named only the second governor in 1843, and whose reforming policies have led him to be regarded kindly by many of today's Māori population—

bought, for a staggering sum, an IBM supercomputer in 2010 to help the country make better sense of the weather and the winds in its vigorous southern latitudes. The government institute dealing with matters meteorological named it the FitzRoy computer. It deals with mathematical concepts and formulae and equations the likes of which Robert FitzRoy would doubtless have had little understanding.

He came from a world, two centuries ago, where wind and weather stood for events and occurrences and phenomena—mesoscale and microscale, to employ the vernacular of the office that he founded—that are intensely physical, can be dangerous or beautiful or memorable or momentous by turn, and that remain in their raw physicality something quite unknown to the arid calculations of today's atmospheric sciences. These are the winds to be found in the pages that follow. Once, that is, these great invisible entities have been measured and analyzed and placed into categories for their strength, their ferocity, and their power, and for the effects they have on the environment over which the winds blow, and on humankind and all his worldly creations.

CHAPTER THREE

Robust and Working Winds

I must go down to the seas again, for the call of the running tide
Is a wild call and a clear call that may not be denied;
And all I ask is a windy day with the white clouds flying,
And the flung spray and the blown spume, and the sea-gulls crying.

—John Masefield, "Sea-Fever" (second stanza), 1902

{ 1 }

Let us remain in Hawaii for a brief moment. Specifically, let us consider the development of the winds on the southeastern coast of the island of Oahu, and consider what happens to the movement of the air on one random spring morning when the sun rises at the start of yet another typically idyllic Hawaiian day.

The first illumination, the first solar gilding, is of the spectacular old lighthouse at the very eastern tip of the island, at Makapu'u Point. Since the Point, garlanded by numberless dangerous islets and reefs and shoals, was invariably the landfall of most ships heading westward across the Pacific Ocean from the ports of California, Oregon, and Washington State, shipowners clamored for a permanent marker to keep their mariners safe and sound. In 1909 the government obliged, and six hundred feet above sea level built a compactly stubby fireplug of a light, just forty-six feet tall and topped with a jaunty coolie-hat roof, that held the largest Fresnel lens of any light in the entire country. It is still a fully functioning lighthouse today—it flashes white twice every twenty seconds, and since its light stands six hundred feet above the ocean, its beams can be seen at least twenty miles away, warning those ships on passage to the Honolulu docks to keep themselves and their fragile hulls well clear of the needle-sharp hazards that litter the waters below the point.

At the top it is almost always windy, usually spectacularly so. Though there have lately been some interruptions to the pattern of the North Pacific trade winds, they seem always to be blowing at full strength up on the Makapu'u lighthouse trail. Occasionally they gust strongly enough to cause walkers and their smaller pets to lurch perilously close to the trail edge and the precipitous drop to the churning ocean below. And such winds are likely to be blowing ferociously at all hours, even at the dawn we are considering here. For trade winds—and prevailing winds generally—tend to blow inside an atmospheric envelope that is maybe a couple of thousand feet thick. This is true just about everywhere, whether deep inland or out on the high seas; there are in such places robust working winds that tend to blow in certain predictable directions at certain times of the year.

Except, that is, in one very particular type of location.

And that unique place, where the notion of prevailing winds does not quite hold true, is around all the world's coastlines. Along coastlines, where sea and land come together, a very different kind of wind caused by a very different set of circumstances takes over from the prevailing winds and locally usurps them as the dominant force of nature. This is where the notion of the *sea breeze* takes center stage. It does here in eastern Oahu, as it does around almost every coastline on the planet.

{2}

A sea breeze is a short-lived thermal wind, universally recognized by any sailor who is working his craft within five or so miles from shore, and its origins are always the same and always have to do with temperature. What takes place as dawn comes up over the beaches below Makapu'u Point takes place also along just about every other of the world's coastlines too, and is eminently simple to explain.

The sun rises and shines down with equally growing strength on the Oahu land surface and the Oahu beaches where the Pacific Ocean abuts the shore. The angle of the sun's rays in Hawaii is very high, this being the tropics, and so in theory whatever lies in the rays' path gets very hot very quickly—but in fact there are significantly different warming effects depending on whether the sun shines on land or on water.

In the sea the water is generally translucent, allowing the sun's light and the solar heating that accompanies it to extend downward to as much as six hundred feet beneath the surface. Twilight is visible down to as much a half a mile, but for the purposes of sea breeze generation, the first six hundred feet are crucial—for this represents a vast amount of water that has to be heated once the sun nudges itself above the eastern horizon. Moreover, the water to be warmed up is in constant motion, being stirred by currents and tides and by the downdraft from its surface waves. All of this makes the seawater very slow to warm up each morning, no matter how fierce the sun or how steep its angle to the sea surface. The result by mid-morning will be water that is perfectly pleasant in which to swim—but when compared to the land against which it washes is almost frigid.

For the basalt cliffs, boulders, rocks, and pebbles and acres of volcanic ash that make up Hawaii's soil become hot very quickly indeed. By mid-morning, a beach pebble that has been fully exposed to the sun will be too hot to touch; beach sand will be uncomfortable to the unshod foot; the soil by the parking lot will be like red velvet pudding; and the only respite for the human visitor will be the cooling waters of the ocean.

The air over the hot land mass will itself now heat up and start to rise, for reasons explained many pages back. And in doing so, the upward-venturing air leaves a region of lower atmospheric pressure above the hot land—low pressure that will attract the cooler, heavier air that is now above the struggling-to-warm-up ocean water, causing it to rush in to take its place, to fill the void. A wind, in other words, is generated: a sea breeze.

This is a wind that is now blowing lustily onshore, soon to become the dominant morning weather feature of the Hawaiian coastline, and it will continue to blow so long as there is a significant temperature difference between the water and the land. And as with all weather patterns, there is little constancy. During the day the two entities will effect a kind of thermal compromise until the moment—mid-afternoon, most usually—when both are more or less equally warm, after which the opposite phenomenon starts to take place: the sun sets, the land cools quickly, and the air above it cools, becoming heavier and sinking downward. The sea, by contrast, dissipates its own painstakingly accumulated heat rather less readily, as if reluctant to do so; high pressure now gathers above the land surface and air starts to move—as wind, of course—from the land to the low-pressure still-risen air mass over the not-yet-cooled sea, and what had been an onshore wind becomes by dusk its kissing cousin, an offshore breeze.

And all this, with its implications for fishing, for transport, for shipping and sailing alike, takes place no matter what is happening up above. For sea breezes operate only up to about a hundred feet, within the narrow envelope that extends maybe three or four miles out to sea. Up above, the trade winds that are steady and strong and a crucial component of what are the world's synoptic weather patterns continue, with majestic aloofness, to blow; one may look up from the Makapu'u beach where one may be firmly in the grip of a cool onshore sea breeze and see evidence—blowing grass, flying clouds, kites, pennants, flags, the whirling anemometer on the lighthouse roof—that fierce northeasterlies are blowing happily up above, having no relationship at all with whatever local mayhem is being created down close to the surface.

{ 3 }

Seven thousand miles from here, and three hundred years before, the young Benjamin Franklin was getting to know sea breezes, and from the occasionally dyspeptic tone of his journal, rather wishing he wasn't.* He was sailing home to Philadelphia in 1726 after two years working as a printer's apprentice in London, and he was a paying passenger aboard the merchant ship *Berkshire*, bound from Gravesend on the Thames to the ports on Chesapeake Bay. Franklin himself boarded at Yarmouth, a small Hampshire port on the Solent, on the Isle of Wight, and for more than a week he and his doughty little ship bounced around in variable winds and breezes of various strengths and directions, the summer morning onshore breezes seeming to try to pinion the vessel in English waters, as if unwilling to let her go. On Friday, August 5, for example, Franklin noted that "towards the night the wind veered to the westward, which put us under ap-

* That Ben Franklin appears here for the second time is no accident; his polymathic abilities and range of interests were such that his presence would be welcome in books about numismatics, typography, philosophy, physics, political discourse, diplomacy, publishing, Latin, and a score of other conceits. He was a far more clever and intellectually well-rounded figure than Thomas Jefferson, and was in later life a keen abolitionist, while Jefferson, even when president, owned hundreds of enslaved men and women.

prehension of being forced into port again [for a third time]; but presently after it fell to a flat calm, and then we had a small breeze that was fair for half an hour, when it was succeeded by a calm again."

And thus did the first week unfold—days peppered with contrary breezes, then with brief periods of fair wind during which the skipper of the little ship could make some progress westward, then a mid-morning breeze that pushed her toward a lee shore in a Dorset bay, until by adroit tacking she extricated herself from what might have been quite a pickle, then fair winds again, then utter calm. "In the afternoon I leaped overboard and swam round the ship to wash myself. Saw several Porpoises this day. About eight o'clock we came to an anchor in forty fathoms water against the tide of flood, somewhere below Portland, and weighed again about eleven, having a small breeze [presumably, given the time of night, a land breeze, allowing the *Berkshire* to shift herself farther offshore, away from the irritating and delaying tar-baby presence of land].

It took Franklin well over a week to make a hundred miles, and until that moment when he was able to write at last, "Took our leave of land this morning. Calms the forepart of the day. In the afternoon a small gale. Saw a grampus." He was off the Lizard in Cornwall, with 3,500 sea miles ahead of him to the Chesapeake entranceway. In 1726 it would have been a long and tedious voyage—made dangerous by a near-total want of navigation marks en route: no lighthouse at the Lizard for another twenty-five years, none at Cape Henry on the far side for sixty-five. It took the *Berkshire* eight further weeks to cross the broad reach of the ocean, until that delicious moment on October 10, 1726, when all the passengers were called on deck to confirm the captain's view that the tree-covered spit of land they could see some miles ahead was most probably Cape Henlopen in what was then Delaware Colony, following which and "to our great joy we saw the pilot-boat come off to us, which was exceeding welcome." He not only led the *Berkshire* through the shoals and perils of the entrance but brought "a peck of apples," which after eight weeks of hardtack and sea biscuits must have tasted like ambrosia.

These days it is sometimes easy to forget what it must have been like to be out in the open ocean and at the mercy of the winds. From late medieval times until the advent of steam power in the mid-nineteenth century, the power of the wind was just about the only means by which seagoing vessels could maneuver and make passage. The centuries before, when large cargo vessels and warships were powered by human rowers—triremes and quinquiremes and then the galleys that were in military service until the Battle of Lepanto in 1571 finally saw them off—did offer a degree of maneuverability that sailing ships were to lack. You could, for instance, move in reverse in a galley—if you flogged the men hard enough to back their oars down. But for long-distance transoceanic travel and for passage through rough waters, man-powered ships were of limited use, as well as being formidably expensive (unless, of course, your crew were impressed slaves, though they still had to be fed to keep them fit enough to row). Once the technology of sail design and the sophistication of the rudder had become sufficiently advanced to take full advantage of the movement of air around the ship, then long-distance trade, exploration, naval warfare—and as a melancholy by-product, colonial expropriation—became both feasible and essentially cost-free, the wind charging nothing for its services. Maybe, as Franklin found out on his tedious westbound transatlantic voyage (eastbound journeys, propelled by a following westerly wind, were then and still are generally faster and easier), obedience to the vagaries of wind has its disadvantages. But for the four centuries before a fully operational steam-powered ship took to the waters, the wind directed everything, the whole world at its mercy.

{4}

"Man hoisted sail before he saddled a horse," wrote *Kon-Tiki*'s Thor Heyerdahl in a later book about oceanic history. "He poled and paddled along rivers and navigated the open seas before he traveled on wheels along a road." Much archaeological evidence—deep-sea fishbones in Swedish middens, Algerian pictographs showing hippopotami being hunted from reed boats, craft made of giant alderwood logs perfectly preserved in Danish peat bogs—supports Heyerdahl's claim, which in any case has the logic of common sense about it.

Five thousand years ago, wandering hunters and gatherers in Mesopotamia not infrequently found their way blocked by rivers—whereupon it took but a scant few moments for them to realize that the fast-moving waters beside which they had stopped, puzzled and bewildered, might in fact do what the desert manifestly could not, and that is offer them a ride, for free. And, if not necessarily in the direction they had planned to go, to a place unknown that might present unimagined new and interesting possibilities.

So, despite knowing nothing about buoyancy, having quite probably no ability to swim, but watching timbers floating by on the water's surface, the braver among their number decided to take a risk and build a primitive boat. The first they made of gathered riverbank reeds or felled trees, using maybe just one log that one wild man among them sat astride, and slid out into the stream. One can imagine the cries of joy, or the terror, or the near-drownings that might well have occurred. But then there would come an hour or a day when the boat remained afloat, stayed upright, turned into something that might be controllable and controlled, and was not merely a thing to be ridden aboard while whirled along on the flood alone. The raging and unharnessed flow of water would take the new vehicle—what one might argue

The triangular lateen, or Latin, sail design here seen on a felucca on the Nile near Luxor. Perhaps the simplest kind of sail, easy to handle by a single sailor, lateen sails are common in the more benign coastal waters of the southern Mediterranean.

was mankind's very *first* vehicle—with it downstream, true. But during this one-way passage it surely occurred to its passenger in short order to imagine he might harness the power of the wind just as well as riding impotently on the water—perhaps using the breeze at first to slow himself, or to steer himself, or even to go against the downstream flow and make his way back upstream instead.

And so he somehow got himself ashore, found himself a straight-enough stick, and mounted it firmly in a vertical position in the middle of his crudely made little craft, making a mast. Then he hitched to this mast the skin of a recently hunted and vanquished animal—a deer, maybe, or an elk, or a gazelle or antelope. Once this skin had been duly mounted, and with his craft now back out in the current again, he would turn the vertical skin this way and that until it caught the wind and billowed and bellied itself out, and the craft suddenly started to move according to this new and, compared to the river water, invisible kind of motive force, and one moreover that, by dint of moving the sail hither and yon, our rider might now control and harness too, allowing in a moment made for anthropological history this boat to evolve into a sailboat, and for the boatman to become the planet's first true sailor.

Rivers provided the first test beds for this new means of wind-fueled propulsion. The Nile most especially—for the simple reason that though the river itself flows from south to north, the prevailing working winds along most of its navigable length blow in almost exactly the opposite direction, from north to south. So an early boat captain might drift effortlessly and languidly with passengers and some cargo for 750 river miles downstream from the First Cataract just outside Aswan all the unobstructed way past a seemingly endless backdrop of temples and pyramids down to Alexandria at the end of the delta and the coast of the Mediterranean Sea. But then, with a single mounted sail and some intelligent use of the sail's potential—by learning, for example, how to tack back and forth across the widest parts of the stream—he might travel with some ease against the current all the way back up to the Cataract, propelled entirely by the wind. Still today the single-masted and lateen-rigged feluccas can be seen plodding gently along the river: to be aboard one such ancient craft with a glass of *karkade*—hibiscus tea—to hand, watching the sun go down over the gold-washed Valley of the Kings at Luxor remains one of life's great experiences, the wind creaking the sheets above you and cooling you as the little craft shifts in time with the currents churning beneath the hull.

{ 5 }

And, of course, most rivers like the Nile end up in the sea,* which means that after testing their sailing skills within the comparative security offered by riverbanks, sailing pioneers might test their abilities and their sails against the challenges of wind and water on the open ocean. In this pioneering phase it was largely up to the sail itself to evolve, the better to make use of the winds that impressed themselves upon it. The profile of the boat's hull would change in time, true; though there is a certain crude simplicity to the design of a floating entity that, unless specifically made for speed or sport or warfare, is mostly offering capacity for cargo, and so is a hollow and sometimes boxy structure that rides on the water's surface while carrying as much as it can hold.

The invention of the rudder—almost certainly by waterborne Chinese travelers—made a vast difference to the navigability of early boats and ships. But the sail is a different matter, with over the centuries innumerable designs and styles and shapes, fashioned of all kinds of materials—animal skins falling quickly out of fashion once weaving began and thus woven materials became available—all seeking to get as much out of the wind as possible.

Warm weather—especially warm water—seems not unreasonably to have been an essential prerequisite for the development of wind-powered watercraft—the prospect of hypothermia being a serious disincentive for a calling that dropped many an early adopter into the sea. So the Polynesians of the South Pacific, and

---

* But not all. Years of greed and theft by farms and cities like Los Angeles have clawed the waters of the 1,500-mile-long Colorado River back from ever emptying properly into the Gulf of California and thereby the Pacific Ocean—the swampy and muddy terminus a sorry lack of gratitude for a river that carved for us such noble natural channels as the Grand Canyon.

the islanders of the south Asian archipelago, the so-called Austronesian inland sea that extends from what is now Luzon south to northern Australia, played a major role in sail development; as did boating communities in the Mediterranean and along the river Nile. The Phoenicians, who in the twelfth century BCE regularly voyaged the two thousand miles between Tyre in the Levant and the Pillars of Hercules, the headlands (Gibraltar being one of them) that guarded the approaches from the western end of the Mediterranean into the Atlantic Ocean, in search of murex shells,* did so in sailing ships. They seem to have used square-rigged sails only—meaning that for uncomplicated forward progress they generally required fair winds, just as Ben Franklin's skipper would later require on the good ship *Berkshire.* The development of sails shaped in such a way, or able to be moved in such a way as to allow the ship itself to move with greater ease forward *against* or at an angle to winds that were not wholly fair—were not blowing in exactly the direction that the ship needed to travel—was thus an evolutionary leap of no little consequence.

Here, as seems necessary, is a brief mathematical excursion, a nod to a field of great interest still—even though it is generally outmoded, save for recreation, as a principal propulsive force—known as *the physics of sail.*

{ 6 }

The more sail, or the more accumulated *area* of sail, the more power; this is generally the rule that sets the tone for getting the

---

* From which, after the removal of a mucus gland through a hole made in the shell, was extracted the valuable blueish liquid that after boiling became known as Tyrian purple, and with which the Mediterranean aristocracy colored their outer clothing—leading to the still familiar phrase "born to the purple" denoting one of noble birth.

most out of any wind that a sailboat skipper might catch. A small boat with a single mast and a tiny canvas sail can go only so fast, can carry only so much weight. But a four-masted barque with five or six enormous sails on each of her masts could positively fly across the sea carrying thousands of tons and hundreds of people. Sail is everything, or used to be.

The precious little oceanic sailing I have done—just a few thousand miles, mainly in the Indian Ocean in the 1980s—was as a very junior crew member aboard a seven-ton schooner that had two masts and heavy canvas sails that were gaff-rigged—meaning that they were trapezoid-shaped and suspended from small battens connected by metal cleats close to the tops of the two masts.

I would estimate at the remove of so many years that the area of each sail—the foremast holding a smaller sail than the main—to be around 150 square feet, say 300 for the two sails combined (or 28 square meters, in the preferred units of the modern sailing world). From this figure one can derive with some arithmetical ease just how much force a wind can press onto the sails, and from that one may calculate just how fast a little yacht like ours might move. The variables in the equation to determine the force, and to be multiplied with the area of sail, include the density of the air through which the boat is intended to move (usually around 1.225 kilograms per cubic meter at sea level), a dimensionless figure known as the lift coefficient that depends on the sail shape and design and on its alignment *at the moment of calculation* to the wind, and then, perhaps most important of all, the speed of the wind itself in meters per second. Multiplying all these variables, one comes up with a number, a value expressed in newtons—the unit being the force that will accelerate one kilogram in the direction of the applied force at the rate of one meter per second per second.

Our little boat had a mass of seven metric tons, or seven thousand kilograms. So, to accelerate such a mass through the water at one meter per second squared would take a force of seven thousand newtons—which, though it sounds like a large number, is next to nothing when compared to the aerodynamic power needed for a suite of sails to move a large cargo ship at, let us say, twenty-five knots. That would require a force of many millions of newtons—a force possibly not attainable by any sails ever made. In our rather trivial case, to bring us up to a speed of five knots, which was our usual cruising speed in a fifteen-knot working northeasterly trade wind, would be about ten thousand newtons, a force quite easily attainable by rigging the sails properly and allowing a fair following wind to carry us on a southwesterly course, the wind firmly behind us pushing us ever onward.

For any given sailing ship, in other words, the speed of the wind and the amount and shape and design and weight of the sails will, in an ideal maritime environment, give the skipper the force he needs to start his vessel moving and then let it run across the sea at a fair lick. But there are different seas and different needs; and while the evolution of early sailing, from voyages up and down the Nile to crisscross journeys through the Mediterranean Sea, was heroic in all senses of the word, it was true transoceanic sailing that had the potential to make the world one, to allow nations to speak peace to all nations, or, of course, and so regrettably often, vice versa. And true transoceanic sailing did not get underway in Mesopotamia or the Levant or along the North African coast or up along its rivers. It began instead in the windy latitudes of the western Pacific Ocean, and while the navigational triumphs of those early blue-water sailors—not least their populating of hundreds upon hundreds of hitherto unpeopled coral islands of what is now generally known as Polynesia—are naturally due in great measure to the design and use of their sails, they also came about because of that uniquely Pacific and Austronesian invention, the outrigger.

Here there is a need for one further but very brief and quasi-technical excursion before looking at the historical importance of the outrigger for long-distance transoceanic sailing. When sailboats sail in high winds, they have a tendency to lean over, to heel. The heeling proclivity depends on the point of sail along which the boat is going, or trying to go, in relation to the wind.

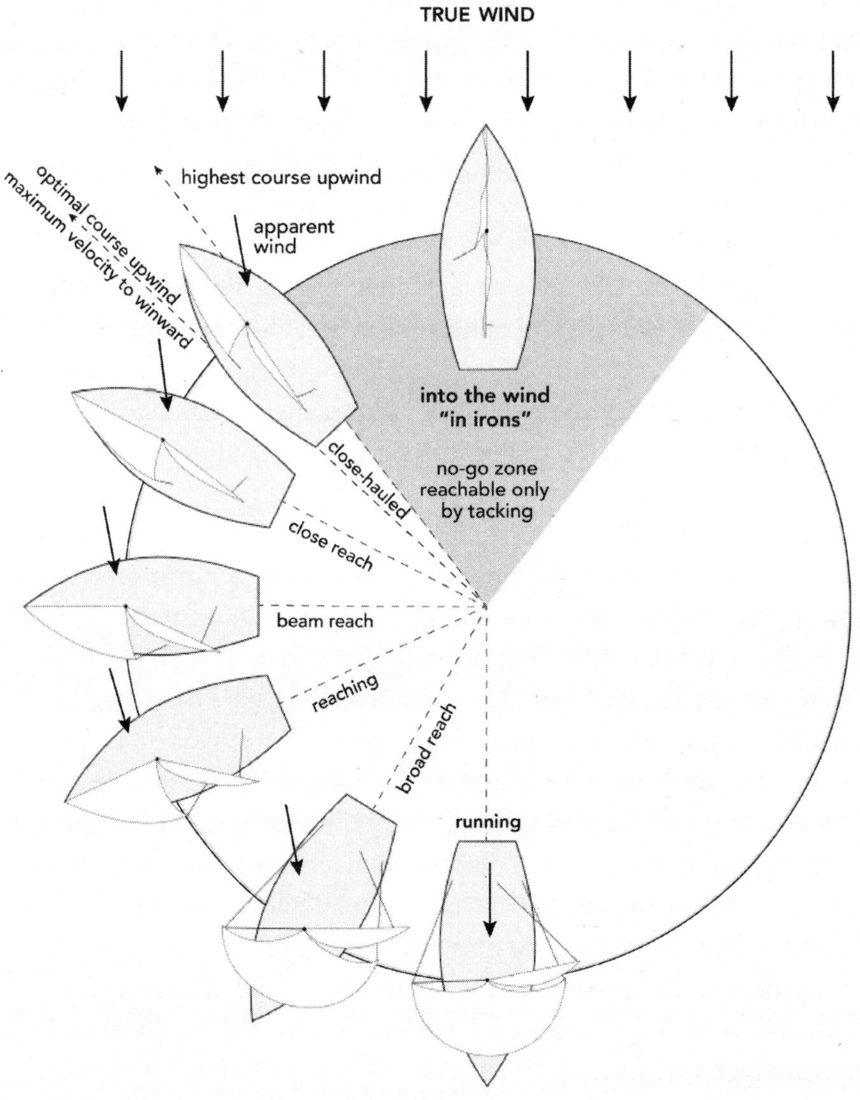

The sailing world has a vocabulary all its own, with the various names given for sailing with or against the wind among the first to be learned by students.

The diagram explains, with our chosen boat the essence of nautical simplicity, having but a single mast and a single sail hoisted up it.

If the boat fully intends to proceed in exactly the same direction as the wind—if, in other words, its sails are full of wind that is coming from precisely astern of the vessel—then the craft is said to be *running with the wind*. It will go almost as fast as the wind, and it will not display any tendency to heel one way or the other.

The very opposite holds as well, though in a rather melancholy way. If the boat is trying to head directly *into* the wind, then, to put it bluntly, it won't be able to. Its sail will flail uselessly, or to use the obvious onomatopoeia, it will *luff*. The boat will not heel in this sorry state either, remaining upright and impotent, a state unwelcome in a variety of other realms.

Just to the left and right of this into-the-wind heading there is about forty degrees of a no-go zone, where a boat will still be quite unable to move ahead. It will be *in irons*, as they say—and the only hope of making progress is for the skipper to ease his craft into that area where the sail is finally able to catch the wind on its edge and let the sail's aerodynamic lift—the same lift as in an airplane, but lifting sideways rather than vertically—is just enough to induce some forward motion. It is here that things start to get interesting, especially if the working wind is a stiff breeze or stronger. For once the boat begins to move, and if the skipper is a good seaman and knows to tighten the sail as tight as can be and command the tiller to head the boat exactly as he wishes, so the craft will be said to be *close-hauled* and will start to heel, to lean over under the pressure of the wind. Sailing thus, sailing *close to the wind*, is, at this juncture, ideal—the boat with its drum-tight sails is leaning twenty to twenty-five degrees from the vertical, and the merest touch on the tiller will either bring it up closer to the vertical once more and slow everything down, or else will heel it over still more riskily into the spray and frighten any newbie crew member, courting disaster and capsize and much embarrassment all around. But this is sailing at its finest—on a close haul in a stiff breeze with the boat doing its aerodynamic best to roar through the water at a speed thought unattainable, particularly when the wind is blowing in what at first blush appears to be a contrary direction.

The other points of sail around the compass perhaps have about them more dignity, more reliability; the boat is more stable—depending, as always, on the skipper. But as one works around the rose and sees the little craft sailing on a beam reach, when the wind is blowing directly across her at right angles, or else when on a broad reach, with the wind blowing about thirty degrees from directly astern—these make for easier, more logical, more commonsensical, more the beginners' sailing lessons; making fine progress, of course, but doing so in a less flamboyant manner, performing sailing that perhaps lacks a little brio and élan. Sailing close to the wind is the dream, or else with the flick of a careless wrist, it can also become the nightmare.

{7}

And the outrigger makes it so much more safe. Which is why its application to the sailing venture of five thousand years ago in the western Pacific is so significant, marks such an inflection point in the manner in which early humankind had long-ago dealings with the wind. So far as employing the winds to move around—to explore, to trade, to spread beliefs, to make war, to seize, to populate—the initial ventures were necessarily hesitant, timid, lacking in true adventure. And often slow: it took ten years for Odysseus to supposedly make the four-hundred-mile passage between Troy and Ithaca, and he was anyway careless enough to have his gifted bag of winds opened by his crew, complicating

Polynesian sailors—master navigators who covered thousands of miles of open Pacific waters—generally employed hull outriggers to give their craft added stability when heeling under the influence of stronger wind.

matters mightily. But true oceangoing high seas adventure—that started not with the somewhat circumspect sailors of the Mediterranean but with the magnificently curious peoples of the western Pacific. And though not directly involved in these passages, it is worth recalling for the record that the aboriginal peoples of Australia represent the oldest settled continuous civilization on the planet, with sixty-five thousand years of cultural continuity, some authorities say. The people of east Asia more generally have long memories, and long-distance sailing, nature-based navigation, and the use of crab-shaped sails and outriggers have long been for them second nature.

It is generally acknowledged by anthropologists today that the earliest sea migrations from the eastern edge of the Asian continent—the area north of the Philippines whose Neolithic inhabitants are known as the Lapita people—took place in large canoes equipped with outriggers. These support floats attached by struts to the side of the hull would keep a heeling boat from capsizing in almost any wind, with the boat going full speed and close-hauled, and yet doing so without any of the perils of heeling at all.

I experienced a little of this feeling of unexpected—but very welcome—stability on an outrigger-equipped canoe off the small Hawaiian island of Lanai, and have not forgotten it. The Belgian sailing enthusiast who managed one of the island hotels instituted as a team-building exercise for his staff a daily predawn canoe race. He would gather the strongest-seeming native Hawaiians that he could find among his employees and divide them into two six-strong paddling-canoe crews, placing them aboard a pair of traditional Polynesian outrigger craft and racing them against one another in early saltwater marathons under the island cliffs. The pair of slender hand-built boats—with their gaily decorated and pencil-thin outriggers attached by great upcurving bamboo struts—raced back and forth through the waves for a good hour or so. Quite often there were perilous swells and unexpected breaks, but the canoe I was aboard slid over them with a dolphin's grace, cutting through the green waters with fantastic speed, unwavering in keeping to its course. The boat kept firmly upright throughout, as the sun rose and it began rapidly to become too warm for comfort. We paddled back to shore through the boiling rips—already sleepy youngsters with their surfboards were gathering to head out to sea to try for the best morning waves—and then we crunched back onto the sand and lugged the fragile-looking, unwieldy-looking craft back up to its resting place under the coconut palms.

The stability that we had enjoyed that Hawaiian morning was a quality that no doubt offered a signal advantage to the Lapita sailors of five thousand years ago. For the past half century archaeologists have been tracking their eastward voyages across the Pacific. They first did so in the 1960s—mainly by excavating some two hundred mid-ocean sites that sport very characteristic

Lapita pottery shards, but also by recording traces of a linguistic commonality that have speared their way between Luzon and Taiwan in the west and Fiji and Tonga five thousand sea miles to the east—passing on the way via Melanesia and Micronesia before settling into the eastern corner of the Polynesian triangle itself. By so doing, and then by having all the work confirmed in the last decade by DNA sampling, these archaeologists and anthropologists have demonstrated the existence of perhaps the most ambitious human migration in history, and all of it powered by wind.

And if not simply by wind, in many cases by *contrary* winds, since in the Pacific Ocean at the latitudes through which the Lapita people sailed, the winds tend to be either northeasterlies (the trades blowing north of the equator) or southeasterlies below, hugely challenging their sailing abilities. But they evidently made it, and did so by designing sailing vessels with outriggers, so that when heading east, and as close-hauled as prudence and safety allowed, they would sail close to the prevailing winds and yet not heel over in a manner that would risk capsize or drowning.

Their achievements are still less recognized today than they should be, despite the gathering evidence of their various landfalls on ever more remote and isolated islands. In 2003 a British explorer named Philip Beale, intrigued by a carving of an outrigger boat that he had seen on a wall at Borobudur, the huge Buddhist temple in central Java, built a replica of the imagined boat—he called it *Samudra Raksa*, or the *Guardian of the Seas*—but then sailed off in her westward, to Africa, which rather misses the point, since there is little evidence of Javanese genes among the inhabitants of present-day Ghana, where he eventually ended up. But he won an Indonesian medal for his trip, which was indeed a heroic affair, even though his wooden craft was equipped with two powerful Chinese outboard motors, just in case.

There is plenty of genetic evidence of travel eastward, however. The fact that there are people today living on islands like Chuuk and Palau in the Carolines, Eniwetok and Bikini in the Marshall Islands, and Tongatapu and Samoa, New Caledonia and the Bismarck Archipelago and Easter Island, and that they all came, essentially, from that region topped, in geographical terms, by northern Luzon and southern Taiwan (and, indeed, in their circuitous routes over the millennia passing by way of such Austronesian sites as Borobudur) speaks of a migration that in terms of distance and difficulty matches at the very least with the much-mythologized achievements of Ferdinand Magellan and Henry the Navigator and Vasco da Gama and John Cabot and the hapless, cruel, misguided Christopher Columbus, and in all probability exceeds them all. And their achievement is not merely the act of getting there; what marks them and their later descendants out is their later ability to discover, chart, and, yes, to populate the ten thousand islands that lie within the Polynesian triangle, the eight hundred thousand square miles of wide-open sea between Hawaii, Rapa Nui, and Aotearoa.

And they did it, this relentlessly curious crisscrossing of the Polynesian seas, in twin-hulled sailing canoes, the second hull a clear evolution from the Lapitan outrigger, and with crab-claw sails and without any navigational aids or equipment whatsoever. Columbus used a compass to get his three little ships to Hispaniola. Magellan had a rudimentary chart, a *portolan*, which nonetheless did not save his life when he confronted an angry chief in the Philippines. But the Polynesian sailors had none of these—just a boat, a pair of sails, and the wind, and a gathering and eventually profound knowledge of the patterns of the sea through which they sailed. It took them a long while—they didn't even reach Aotearoa, the *land of the long white cloud* that we Westerners who high-handedly and wrongly believe we discovered call New Zealand, until some seven centuries ago. But they got there with only the force of the working winds to pull them through the seas, a feat that quite astonished Captain James Cook, who saw what they had done, and how they did it, four centuries later.

{8}

Three centuries later still a group of Hawaiians did it again. As the state's official bicentennial tribute to the United States, the Polynesian Voyaging Society built a very large full-scale model of a traditional double-hulled sailing canoe, the kind that Captain Cook had seen and drawn in his logbook centuries before. Her name was *Hokule'a*, translated as *Star of Joy*, and though she herself was seen as a magnificent craft, a triumph of craftsmanship, there was no little controversy about her being the chosen gift. After all, Hawaii's modern American history exposes a litany of misery. In the late nineteenth century a cabal of wealthy white mainland businessmen had engineered the annexation of the Hawaiian Kingdom, overthrown the monarchy, and used US Marines to arrest the reigning Queen Lili'uokalani, forcing her to abdicate and leaving her to remain ignominiously in Honolulu living as a private citizen until she died, a tragic and harshly treated figure, in 1917. So, to this day, and understandably, many native Hawaiians—most, some surveys say—remain bitterly opposed to the *haole* business community that ran so roughshod over the ancient island society and remain profoundly hostile to the idea that a sailing canoe, so symbolic of Hawaii's particularly Polynesian pride, should be offered as a birthday gift.

Nonetheless, the sixty-foot-long vessel has in the half century of her working life so far proved a triumph, and on many levels has worked wonders. Her sailing achievements have rekindled pride, especially among young Hawaiians, in their Polynesian identity. And that is because she has undertaken a very long transpacific journey—initially the two thousand miles to Tahiti—just as the Polynesians of the past are known to have done, with just the ocean swells, the passing birds and fish and sea mammals, the sun and moon and stars and planets—and the wind—to guide her on her way.

The vessel made her first test voyage under such conditions

and constraints in 1976, with an elder, a traditional navigator from the Caroline Islands, to offer guidance and encouragement. The crew had no compass, no radio, no sextant, no charts, not even wristwatches from which they could derive direction. Yet they made it to the very island off Tahiti for which they had aimed, and arrived exactly on time, such that thousands of fellow Polynesians were on the Papeete quayside to greet them and savor the shared delight of a pan-oceanic achievement. Since then, the craft has sailed tens of thousands of sea miles, some of the journeys intended as expressions of amity, others as instructive expeditions to help young Hawaiians find greater purpose in life. Most impressive of all was a three-year circumnavigation of the world, beginning in 2014, in which for the first time *Hokule'a* left her comfort zone of the Pacific Ocean—where the patterns of constellations had long been firmly impressed into the navigators' minds—and ventured north of Australia into first the Indian and then the Atlantic Oceans. But she sailed impeccably across forty thousand miles of sea, underlining for all the world to see the enduring legacy of traditional sailing methods.

Except, at the end of this examination of the significance of outriggers, there was an event that prompts us to pause and note a single and not insignificant caveat. Early on in her otherwise estimable career, *Hokule'a* capsized. The supposed stability offered by the Lapita-style outrigger—which had served all too well the greatest navigating communities the world had ever seen—failed the crew of the newly made vessel. And what is more, it did so fatally.

It was March 1978, and the canoe was heading out on another southbound passage to the Friendly Islands, with Tahiti once more the chosen destination. There were sixteen crew members aboard, bent on a slightly unusual mission. Together with the traditional navigators, who were pledged as usual to employ no aids whatsoever in their voyage, there were two crew members who did carry instruments and who would check the accuracy of the boat's assumed position day by day to test whether traditional methods worked on the short-haul sectors of the journey, rather than just evaluating the success of the voyage overall. They were similarly pledged not to intervene unless the craft was in imminent danger of hitting something or foundering.

They left just as darkness was falling, from a harbor to the east of Waikiki. Once clear of the moles they were met, as expected, with the full force of thirty-knot following trade winds. The seas were moderate, though with swells of up to ten feet from the northeast, typical accumulations from sustained trade winds and, again, not unexpected. The forecast was for moderating weather, and the captain had deliberately left when he did to take advantage of this fairly robust weather, familiar in the canoe's previous experiences and likely to get the voyage off to a fast start.

But five hours out, everything went wrong. The two hulls were crammed with food for the monthlong voyage and were very heavy. The starboard hull, which was to leeward in this situation of a southbound boat in a northeasterly wind, dipped beneath the water repeatedly and, so far as a later inquiry showed, seawater got over the coaming and into the hatches. The boat became ever more cumbersome and difficult to steer and to handle. Shortly before midnight the sails swiveled the lighter windward port-side hull such that it moved up and around the sunken leeward hull, which was now essentially immobilized and dead in the water, and the entire craft flipped in one long, slow arching catastrophe. Though the lights of Honolulu and Molokai and Lanai—where I would have my own outrigger epiphany some decades later—were visible, twinkling merrily, and all Hawaii was alive and well around them, the sorry, sodden boat and its shivering crew clinging to her upturned hulls were as isolated and incommunicado—for they had no radio, traditional sailing being the expedition's watchword—as if they were castaways up in the Northwest Passage.

When dawn broke, one of the crew, a young but highly experienced lifesaver named Eddie Aikau, decided to untie a surfboard and paddle to the nearest land for help. The crew watched as he rode the waves northward. He was never seen again.

It was another full day of mounting panic for the survivors as their craft bobbed slowly westward on the currents, drifting helplessly farther and farther away from the network of inter-island airline routes with which all Hawaiian airspace is traced. No one ashore had evidently listed the craft as missing. No passing ship could see her, lying so low, with her normally visible and distinctive masts and sails underwater. They watched the aircraft taking off from the nearest island airports and fired Very lights and flares to alert them, but without effect. Until, finally, shortly after 10:00 p.m. on their second night on the wreck, they saw the last Hawaiian Airlines flight of the night take off from Kona, bound for Honolulu, and watched it take a slightly different and more westerly route than usual. They fired volley after volley of flares, and the pilot spotted them, realized there was down there in the Stygian gloom a vessel in distress, circled low over the *Hokule'a*, and flashed his landing lights to let the much-relieved crew know they had been seen. He radioed the coordinates. A Coast Guard helicopter was hovering overhead by midnight, a fast cutter beside the stricken craft by two in the morning, and as the sun rose on March 18, a tow rope was tied to the bow of the now re-righted canoe—a rogue wave had tipped the boat over once again—and the sad and sodden and undignified assemblage was returned to port, with one of their number, Eddie Aikau, never to be found. His surfboard was later spotted and recovered, but nothing else.

As a consequence of the accident, *Hokule'a*, from then on, was always to be accompanied on its longer voyages. A chase boat, standing a respectful distance of a couple of miles astern, was always to be there, just in case, most particularly on the 2014 circumnavigation. Crucially, this following boat, though having the appearance of tradition and antiquity, was not averse to keeping high-tech wizardry discreetly hidden from view. It had a radio. Gyrocompasses. Up-to-date charts. And GPS.

Its constant presence as a blue-water babysitter not surprisingly irked the purists of the Voyaging Society, even as the 1978 accident reinforced the view that still holds today: that the Lapita sailors of old—who voyaged alone or in small groups, without chase boats or radios or compasses or anything to help, save for their sails and their precious flimsy little outriggers—accomplished something quite unimaginable. They had harnessed the ocean winds in a manner few of their successors, in any ocean anywhere, would ever be able to equal.

{9}

In those very early years of human existence, few such sailors inhabited the world's other great oceans. Perhaps it was sheer terror of the forces apparent on the sea's edges that kept them away for so long. Perhaps the Asian canoeists who set out from Lapita and Austronesia felt somehow that they had an easier go of it by heading eastward, lured into the deep by calmer waters and fairer winds, since they were heading, after all, in the selfsame direction that planet Earth was rotating—not, of course, that they knew any of this, beyond the sailors' hunch. But it was so different on the fringes of the great gray Atlantic Ocean, however. Some other force kept them landbound for much longer.

For a Viking longboat captain riding safely on the still waters of his home-port fjord, hemmed in from the gales by the immense protective granite cliffs; for an Irish missionary with his faithful oarsmen standing by in the shelter of Bantry Bay in southwest Ireland; for a Phoenician skipper gazing, hungry for yet more valuable murex shells that he rightly surmised might lie in abundance out past the great mountains of Tunisia that rose on his left and Andalusia on his right—to all these early mariners, the waters and the winds that lay ahead must have seemed intimidating, horrifying, without so much of the lure that tempted those far away and of whom they had no knowledge, and who had yet launched themselves off into their windswept and watery unknowns thousands of years before. It is idle to wonder why it took the Europeans and Levanters so long; but eventually they did indeed screw their various courage to the various sticking places, did ease their wooden ships off the gravel of their beaches, did haul their sails—square rigs all—up their masts, and with prayers and nail-biting moments of deep anxiety, did set off westward, into the maelstrom. To most, and at least for a while in all their voyages, the winds into which they set sail were of a decidedly contrary nature, blowing against them just as would the sea breezes that so irritated and delayed Ben Franklin centuries later, making it hard to fight the spume and the swells. The seas were high and the waters were cold. But these men—Irish, Norwegian, Lebanese we would term them now—persisted, and ventured out time after time until their doggedness bore fruit and, one by one, their little expeditions found fairer winds and easier seas, and one after another, they each made it.

As to which winds blew whom the fairest, much amiable dispute continues. Admirers of St. Brendan, the sixth-century Irish Christian abbot who devoted much of his life to spreading the Gospel throughout the west of Ireland, hold still that he and a crew of muscular monks successfully crossed the Atlantic—and returned home safely—a good eight centuries before Columbus. The dramatic voyage is chronicled in a tenth-century Latin manuscript filled with episodes of great excitement and danger, and with charmingly fanciful illustrations of the saint standing to give sermons so powerful that fish would surface around the little curragh to listen with awed respect.

The enterprising and commercially savvy British explorer-historian Tim Severin, armed with a generous publishing advance, replicated St. Brendan's voyage in 1976, first building by hand a thirty-six-foot version, in ash and oak lashed together with miles of leather thongs and wrapped in half-inch-thick animal hides, of a two-masted curragh, similar in all respects to the tenth-century manuscript images. With a crew of three—and with a radio, compass, flares, and life rafts—Severin set off in May and voyaged first northwest to the Scottish Hebrides, then up into the fogs of the Faroe Islands, and finally to Iceland, where they let their barnacled but otherwise sound little boat rest through the winter.

The following spring the crew flew back to Reykjavík to rejoin the boat and crossed the Demark Strait in the hope of seeing Greenland, but ice prevented them from getting close, and compelled them instead to head southwestward—which they did, dodging the kind of icebergs that had sunk the *Titanic* sixty-four years before, until they succeeded in making the small fishing community of Peckford Island, off the northeast coast of Newfoundland, Canada. They had traveled 4,500 miles, and their handmade flax sails, square-rigged and suspended from two masts, had given them precious little trouble. On occasion the crew had used oars—just as St. Brendan's monks reportedly had too—but generally it was the steady and reliable southeasterlies and then northerlies that blew them along their track, convincing Tim Severin, when finally he returned home to Ireland and deserved adulation and the detailed inquiries of anthropologists and historians, that Brendan could very well have achieved what he and his men had now shown was possible. The well-known Irish legends, the *immrama*, could well have been based on fact. But no evidence had ever turned up in Eastern Canada to suggest that a group of Irish monks had arrived there or, indeed, anywhere in North America, and so the story has ever since remained a charming mystery. The jury is still out.

{ 10 }

As it is for whether the Phoenicians should have taken the laurels. Thirty years after Severin's Atlantic crossing, Philip Beale, who already had some experience with an Indian Ocean crossing in his replica of a Javanese sailing craft, decided to test the little-held hypothesis that the genius sailors of the Mediterranean might indeed have landed on American shores. As captain of the Syrian hand-built single-masted *Phoenicia*, which he had already tested by a circumnavigation of the African continent, he set out from northern Tunisia in 2019, put in at Gibraltar and Cádiz, headed a little way southward to Essaouira on the Moroccan Atlantic coast, and then slid across to the Canary Islands and Tenerife before setting his purple-and-white-striped single sail for the long trek westward. It took him and his decidedly international fifteen-strong crew thirty-nine days to reach the Americas, at the very eastern tip of the Dominican Republic, taking almost exactly the same time (and landing in almost exactly the same place) as Columbus had done back in 1492.

Captain Beale then went on to Fort Lauderdale in Florida before something quite unusual took place. The Church of Jesus Christ of Latter-Day Saints—the Mormons—took Captain Beale to be in some sense a reincarnation of the prophet Lehi, who according to the *Book of Mormon* left Jerusalem and the Levantine coast to sail his own way to the Promised Land—the land where, eventually, Brigham Young supposedly declared "This is the place" and founded Salt Lake City in Utah. A puzzled and flattered Captain Beale happily agreed that the Mormons could transport his boat and all its accoutrements to a brand-new Phoenician Ship Museum on the left bank of the Mississippi River at Fort Madison, Iowa. Which is where it remains to this day, an unexpectedly sited Midwestern memorial to a pair of Levantine wind-driven expeditions—the one quite recent and involving a middle-aged Englishman and his replica of an ancient sailboat (though, like his Indonesian-made predecessor, equipped with engines, radar, compasses, radio, and automatic life rafts), the other fanciful, performed many centuries ago, led by a sailor-prophet named Lehi who, with his wife, Sariah, their children Laman,

Lemuel, Sam, Nephi, Jacob, Joseph, and an assortment of daughters, made much the same transoceanic journey carrying with him the precious golden plates that the teenaged farm boy Joseph Smith would later discover buried in a hill in a far corner of upstate New York, and following his translation of the reformed Egyptian language scribed on them founded a religion in 1830 that now has some seventeen million members and thousands of congregations and even more missionaries—many working, as it happens, in Polynesia.

Though there is no suggestion that the young men who are to this day still spreading the word of Mormon in the South Pacific ever reached their destinations by outrigger-equipped sailing canoes, it is perhaps worth noting the direction of the winds when Prophet Lehi launched himself off into the waters of the Atlantic six centuries before Christ. Had the winds not blown fair and propelled him and later his son Nephi to sequester his precious plates in upstate New York, they never would have been unearthed by Mr. Smith, nor would Mr. Young ever have taken his Conestoga wagon across to the Salt Lake Valley, and the Church of Jesus Christ of Latter-Day Saints would have been perhaps a very different animal, possibly based in an entirely different city, maybe even in a very different country.

{ 11 }

The winds blew with much more certainty for the Vikings, however—for they did indeed sail westward, and did not need any modern-day replica craft to support their claim for having made landfall in the Americas. They left their own memorial, found by chance in 1960 when a local Newfoundland farmer advised a pair of passing Norwegian archaeologists to investigate a cluster of grassy hillocks outside his village of L'Anse aux Meadows. The pair, the husband-and-wife team of Helge and Anne

Ingstad, had come to this distant corner of far northern Newfoundland to investigate suggestions made in the Icelandic sagas of the existence of a place called Vinland. The Vinland Map, only recently declared by its owners, the Beinecke Rare Book & Manuscript Library at Yale University, to be a twentieth-century fake, was being spoken of in 1960, and the Ingstads, captivated by all matters Vinland, were eager for any constructive suggestions as to where Norse sailors might have landed, as the map declares they did. So, in the summer of 1961 they began a half-decade of digging at the grassy mounds, suspecting them at first to be the remains of a settlement of local aboriginal peoples only. Their eventual discovery and realization that this was in fact a large and sophisticated *Norse* settlement has since placed them, and the eight buildings they found, in the pantheon of world archaeology's most venerated.

The eight low, timber-framed and sod-covered little buildings turned out to be filled with Norse items and implements, showing incontrovertibly that it was Norsemen and women, not Genoese sailors, who first settled in North America. And of course the word today is very much "settled," not "discovered," since Newfoundland already had, and for eight thousand years prior, its own native population scattered across the island.

L'Anse aux Meadows—the Bay of the Jellyfish to most—is now a fully fledged World Heritage Site, looked after by Parks Canada. It consists of three dwellings, a forge, and four workshops, and all the uncovered artifacts suggest that those who came smelted and forged iron, performed heroic acts of woodworking, and repaired ships, as well as carrying out all the domestic chores associated with long-term settlement.

The widespread assumption among Norse scholars back in Oslo and Bergen is that the principal settler was, just as related in Iceland's *Vinland Saga*, Leif Erikson, the famous son of the even more famous explorer of Greenland and Iceland, Erik the Red. Leif is said to have sailed to Newfoundland from the southern tip of Greenland, about seven hundred iceberg-dodging miles across the Labrador Sea, in 1001 CE, with most carbon dating suggesting a little later, around the first decade of the eleventh century. It is also said—fancifully, most probably, though it is pleasing to imagine it to be true—that in 1004 a boy was born here, or maybe a little farther south in Vinland proper—and was named Snorri Thorfinnsson.

If true, then Snorri was the first European to have been born in North America, and almost half a millennium before Columbus supposedly fetched up on the sands of Hispaniola. As it happens, Snorri and his parents, shipmates and friends of Leif Erikson, stayed only a very few years, and then, citing problems with the native Newfoundlanders whom they rudely termed *Skraelingar*, barbarians, returned to the relative safety and comfort of Iceland, where they settled down for good. Snorri is known to have fathered two sons, after which the tale of this unique figure in world history vanishes into the cold northern fogs.

Though there may have been no pressing need for any present-day Norwegians to prove and confirm the maritime achievements of their ancestors, many modern adventurers have done so; and in all cases one thing stands out: the elegant beauty of their boats. The Irish curragh and the Phoenician sailing ship have a rough-and-ready appearance, possibly well able to do the job, but they lack the sleek elegance of a wooden boat that was crafted in Scandinavia a thousand years ago. It is of course unlikely that those Parisians who watched a flotilla of longships coming up the Seine in 845 CE—as they did again twice more, before laying full-on siege to the city forty years later—considered the beauty of their assailants' watercraft, though doubtless they would later admit ruefully to the awful precision of the raids. And it was this pan-Scandinavian combination—of Georg Jensen design and Volvo efficiency—that helped make the Vikings so effective a maritime people. They may at first have been hesitant to leave the shelter of their immediate coastline, but once out on the open waters, they came rapidly to know how best to sail, to take full advantage of the cold North Atlantic winds, and how to travel farther and faster than anyone else from mainland Europe—and, moreover, to leave as a World Heritage Site their mark where they eventually settled.*

{ 12 }

And almost all of these early craft were powered best with square-rigged sails. Less ambitious journeying, and most sporting ventures, throughout history and still today, more usually employed fore-and-aft rigged vessels, in which the sails were set along the line of the ship's keel, or along the longest line from stem to stern. Most pleasure yachts one sees today are, like the gaff-rigged schooner on which I sailed the Indian Ocean, rigged thus. But not for cargo, nor for war, in the Atlantic and then in the seas beyond. There, all was square-rigged. Some had supplementary oars, of course, for fussing around in windless bays and estuaries and once in the shelter of ports. But out at sea, the wind alone was the source of propulsion, and large square sails were undoubtedly and according to the immutable laws of phys-

---

* Sweden, the coastline of which is not truly exposed to the open ocean—the Kattegat and the Baltic coasts are fairly well sheltered—has a somewhat less than happy public relationship with the wind. Stockholm's most popular museum houses the seventeenth-century three-masted sailing ship the *Vasa*, which in August 1628 turned turtle and sank less than a mile from where she had been launched a short while before. At the time she had four sails up and was moving well in a light breeze, an audience of thousands of excited Swedes on hand to watch her sail away to join a local war. But a sudden gust of westerly wind caught her unexpectedly, she heeled to port, and water cascaded in through the gun decks, dooming her to sink in one hundred feet of water, two of her three masts remaining above the waves as memorial to spectacularly poor seamanship. She was raised in 1961, found to be remarkably well preserved, and stands today as the counterpoint to Norway's sailing success three thousand miles away in North America.

ics the best means of capturing it. Which led—initially in the Atlantic Ocean where they were born, and then to the world's seas beyond—to the near-universal adoption of square rigging for the subsequent eight centuries (or four, if Columbus is seen as the starting point) of what is now generally known as the Age of Sail.

This age—overlapping with those other historians' inventions like the Age of Exploration, the Age of Discovery, and the Age of Empire—was by its very nature a creature that depended on the wind and on how well or ill the succession of ships and their crews could handle it.

During those centuries when sailing vessels dominated the oceans and the lesser seas, voyages became longer and more perilous, naval battles became more frequent and the goals of their wars ever more ambitious, the cargoes to be carried outbound and the expropriated plunder to be brought home—human cargo included, during the Age of Slaving—became ever more massive, and trade, that all-important lifeblood of an ever-enriching world, developed near exponentially, and competition between traders became ever more fierce—and yet all this took place while the winds just blew and blew as they always did and—subject to the Great Stilling caveats already noted—probably always will. Nothing about the wind changed significantly during those centuries; it was therefore up to the sailors to hone and refine their seaborne skills, for ship designers to refine their designs, and for those who made and employed sails to create them and deploy them in those myriad ways that satisfied the needs of the various ships and their purposes.

They started small. The best-known ships of the day—the *Niña* and the *Pinta* and the *Santa Maria* of the first Columbus venture, the *Matthew* of John Cabot's Tudor voyage to Newfoundland, and Amerigo Vespucci's journeys to what is now South America—were all caravels of less than sixty feet long and weighing in at maybe fifty deadweight tons. The dainty caravels were tasked with exploring and needed to be both nimble and durable. A century and a quarter later, in 1620, the *Mayflower*, a craft not bent to exploring but rather to the transport of settlers to what by now had already been explored, was rather larger—one hundred feet from bowsprit to spanker boom, and of 180 tons burthen. She crossed the sea with 102 Pilgrim settlers, with two dogs and an assortment of other farm animals, from chickens to cows, and a crew of 30 under the captaincy of Christopher Jones. They experienced much the same frustration with persistently unhelpful sea breezes off the southern English coast as Benjamin Franklin would endure the following century, when he was trying to beat back from London to Philadelphia; and en route there were fierce November storms that nearly wrecked the little vessel and caused at least one child to die (and yet allowed another to be born).

Contrary winds then played havoc with their landfall. Captain Jones had planned to head for Virginia; when storms nixed that plan, the ship headed for the mouth of the Hudson—except that, two days out, the wind once again changed, veering from easterly to southeasterly, and the hapless skipper, having most of his sails set well-nigh immovably, had little choice but to let the cold late-autumn storms blow him ashore some scores of miles to the north of where he had intended. Plymouth Bay Colony was to be the settlers' centerpiece in what would become the colony of Massachusetts, when by rights it might have been some miles elsewhere—another small example of how the fickleness of winds (see Chernobyl) can have unanticipated consequences.

The *Mayflower*, though built in eastern England, was of a common Dutch design known as a *fluyt*. As such she was one of scores of designs created around the basic idea of a square-rigged sail set, and which, as the years wore on, became ever larger and sturdier and magnificent, and still more confident in dealing with any distances, sea states, and, most important, wind. Since research of the time demonstrated that the winds became more constant and predictable the higher they blew above the water—down at sea level wind is interrupted and deflected by the very waves it helps create, among other things—there was a premium on very tall masts, with tightly tensioned stays to keep the masts upright and triangular sails suspended from these stays, and with square sails suspended from yards ranged from the very top of each mast with as many as six or seven similarly shaped square sails, increasing in size the closer they were to the deck.

As the sails grew larger, the ships got ever larger.

There were fluyts and snows; then caravels and galleons, clippers and barks, brigantines and barquentines, and ships with three masts, four masts, even five. There were topsails, skysails, topgallants, royals, courses, spankers, jibs, staysails, and spinnakers, all named according to the mast from which they were suspended so a captain might order, if a squall was seen approaching, the reefing of a lower fore topgallant or, in order to wear the ship and sail her with the wind, the bracing of a lower main topsail. To a newly signed sailor, or maybe one shanghaied or impressed by cunning or force in a dockside bar and thus put to shipboard work unwillingly, such would be a near-impossible learning curve to master. For a landlubber with just a vague awareness of how a breeze might riffle the grass on a nearby meadow suddenly to be required to know how best to take advantage of the subtleties and fickleness of the infinitely more complex movements of the air at sea—such that if you get it wrong and heave on this sheet instead of that one, the cat-o'-nine-tails awaits, or, if the bosun was a terror, the certain lethality of being keelhauled. Small wonder that single men drinking in those eighteenth-century dockside bars took precautions to ward off the clever tactics of the press-gangs.*

* He would be extremely friendly, this navy recruiter, and would readily stand you a pint of ale, but as you were drinking would slip a shilling coin into your beer such

Three quite distinct functions determined the kind of vessels that were ordered from the shipyards of seagoing countries. There were cargo carriers—capacious, strong, fast, reliable, solid. There were naval vessels, armed with batteries of cannon—the vessels needing to be light, highly maneuverable, fast, and repairable. And there were whaling ships, which could be slow and steady, easy enough to sail with small crews, and, since the actual whaling was performed from the longboats that the big ships carried, having something of a motherly quality about them, a floating home for the sailors who went out and did battle with the great mammals it was their business to slaughter.

{ 13 }

The good ship *Peking* was until 2017 moored alongside a pier in Lower Manhattan, the most impressive attraction of the seven ships floating beside the city's then estimable seaport museum. Though she was Hamburg-built, she had been in New York for forty years, her massive seventeen-story masts and double yards a familiar sight between the skyscrapers, with the East River and the Brooklyn waterfront a fitting backdrop for a ship that for much of the last century had worked up and down and across the Atlantic. But by 2017 she sat silent and somewhat forlorn in a section of the city that was changing all too rapidly around her—the Fulton Street Fish Market closing down in 2005, Tropical Storm Sandy doing immense local damage in 2012, and the publication of long-term plans that didn't appear to include her.

---

that when you drank the tankard dry you would be deemed to have accepted the King's Shilling and thus be officially under government contract to set sail as your kindly host demanded. Innkeepers, according to a tale believed by the credulous, had tankards made with glass bottoms, so that a possible victim might spy the dreaded coin and turn down the offer.

To save money, the museum eventually decided to pension off the century-old ship, and had originally planned to send her to the breaker's yard to be scrapped. But the German Port Museum back in Hamburg, where she had been built in 1911, stepped in and agreed to take her off New York's hands; and so in the late summer of 2017 a huge semi-submersible ship stopped by, lowered herself down into the sea, and allowed the 2,800-ton *Peking* to slide in between her twin hulls, refloated herself with the sailing ship now safely drydocked inside her, made an easy sixteen-day summertime passage across the ocean, and deposited her back where she had been born, a century and a bit before. She is now the gleaming and refinished centerpiece of a museum dedicated to reminding visitors of the maritime history of a once-great trading nation, and most especially the way in which huge sailing vessels like this—the last of their kind, before steam power took over at the start of the twentieth century—managed the most dangerous route of all, sailing westbound—the wrong way—around Cape Horn.

The *Peking* was typically dedicated to sailing this so-called Valparaiso Route around the dreaded Cape, regularly hauling about five thousand tons of European manufactured goods down to Chile and bringing a similar tonnage of bagged fertilizer—usually bird guano—back to Bremen, from where she had started.

The route's first few hundred miles were fairly unexceptional. She would first head out west into the North Sea, then turn south, hoping for decent weather in this notoriously rough stretch of sea, then squeeze through the ship-crowded English Channel until, after three hundred miles on a westerly heading, reaching and turning southward into the open Atlantic after passing Cornwall's Isles of Scilly. After that it would be mostly plain sailing, as the phrase has it: rounding the North Atlantic High, battling the westerlies (with many gales in January, precious few in July), picking up the northeast trade winds once south of the Azores, heading as best as the doldrums allowed across the equatorial latitudes, next making such use as possible of the southeasterly trades off the Brazilian coast, and then standing by to watch for the daunting challenges ahead.

This would mark the end of the more mundane part of the sailing. From here on, south of forty degrees, the winds would steadily start to change, with ever-strengthening westerlies ahead—the Roaring Forties—and many gales all year round, though maybe with just a little lessening of their ferocity and frequency in July. From here on the perils would be very considerable, on the approach to and then in the very act of rounding Cape Horn—or, as the navigational handbooks term it by its proper Chilean-Spanish name, *Cabo de Hornos.*

{ 14 }

The definitive navigational handbook is a volume that I have owned for fifty years and from which I would be loath ever to part. It is published by the British Admiralty, is titled *Ocean Passages for the World*, and gives well-founded advice to mariners on how to get from almost anywhere to almost anywhere else by sea, over the safest and most economically sensible route. Need to make passage from *Yokohama to Panama*, from *Montreal to Gibraltar avoiding the shoals of the former cod fisheries of the Grand Banks*, or from the *English Channel to Cape of Good Hope*, and you will be advised how best to do so.

In this last case, however, there is one telling line in what turns out to be the rather more than usually involved instructions for how to get from Dover to Cape Town. It begins thus: "Having crossed the equator, stand across the South East Trades on the port tack . . . for the winds will draw more to the East as the vessel advances . . ."

The reason for this is because this particular instruction, unlike the two previous, comes from part two of the volume, which is all about the recommended routes for *Sailing Vessels.*

The bulk of the book deals with how *Motor Vessels* might plan their routes—which will clearly be very much easier, since a motor vessel can go more or less anywhere she pleases, whatever the weather, whatever the direction of the winds. But a sailing vessel—like the *Peking*, the square-rigger whose exploits we are presently considering—has to be much more prudent, given the vagaries of sea and storm, of current, of waves and weather. And of wind, from which she wins her propulsive force. Perhaps nowhere in the world is the need for caution, advice, and prudence more pressing than on the approaches to Cape Horn, Cabo de Hornos.

The following, lightly edited, is what the master of the *Peking* would read, having plucked the dark-blue volume of Admiralty advice down from the shelf above the chart table.

> **From the Atlantic to the Pacific—Rounding Cabo de Hornos westbound.** The usual track is to take as direct a course as possible from a position 200 miles east of Rio de Janeiro to . . . pass 30 or 40 miles east of Isla de Los Estados . . . there is off its easterly extremity a heavy tide rip which extends for a distance of 5 or 6 miles or even more to seaward. When the wind is strong and opposed to the tidal stream the overfalls are overwhelming and very dangerous even to a large and well found vessel. Seamen must use every precaution to avoid this perilous area.
>
> But the Estrecho de la Maire provides the shortest route around Cabo de Hornos with a valuable saving when the difficulty of making westing is considered. The conditions must be suitable. A passage of Estrecho de la Maire is best attempted during daylight with the fair wind and tide the best time for beginning the passage through being at one hour after high water. A vessel should if necessary heave to at the entrance to the Strait until that moment. Under these conditions even should the wind fail or come adverse a vessel would probably drive through rapidly for the tidal streams are strong. With a southerly wind it would not be advisable to attempt the Strait for with a weather-going tide the sea is very turbulent and

> might severely endanger the safety of a small vessel and do much damage to a large one . . . with snow and ice occurring at about the time of the equinox . . .

{ 15 }

On this barque's four masts are a total of thirty-eight sails, the larger courses and mainsails weighing in at a ton apiece—very much heavier when wet or frozen solid with sea ice. All told, the sails present a full acre of canvas to the wind—and when you consider that the cleared acreage on which the Liljestrand House back in Hawaii sits, discussed in the previous chapter, is just half an acre, accommodating the main house, the gardens, the swim-

The iron-hulled four-masted 3,100-ton barque Peking, *launched in 1911 for the Chilean guano trade. After retirement—and spending time in the UK as war reparations—she became a museum ship in lower Manhattan and is now back as a floating memorial where she was built, in Germany.*

ming pools, and a scattering of outbuildings contained in an area that is just half the sail area of the *Peking*, then it becomes possible to imagine what force the wind can exert, what power the wind can bring to bear, on the ship. During one passage in the trade winds a glimpse over the stern rail shows the sea streaming by, leaving, with a half-mile-long wake of white, the water so boiling and so Niagara-like that one could swear four bronze propellors are whirling and thrashing away beneath the stern. But no—aboard the *Peking* there is not a single device powered by steam or electricity or by any force other than wind or human muscle. Even the lights are kerosene, and the immense winches that bring down the *three hundred* different ropes and sheets and stays and braces that can lift or lower the sails or rotate the yards around the masts to tack the ship in times of aeolian stress are all moved by the crew, who emerge from a journey like this—if they emerge at all; and on one passage in 1929 from which there exists a diary, two who were swept into the ocean and never found did not—unimaginably strong in hand and spirit.

Imagine, if you will, the moment when you are finally through the Strait and done with that *other* Staten Island and all the land, white-crested mountains and jagged cliffs and one single lighthouse perched on the edge of a skerry, is now piled up on your starboard side—all South America is to starboard, the whole inhabited world is to your right—so even a common seaman holding on for dear life on the monstrously swaying deck knows that this is it, this is the moment when you are off the Horn, Cabo de Hornos, and the seas are piling up and the wind is raging from due west, dead ahead, howling from the very direction in which you are now trying to push, smashing directly into your bow, which plunges into the waves, diving twenty feet under the green-and-white madness that is the place where the two oceans collide with an unyielding forty-, fifty-, or sixty-knot gale, and then you hear through the din a shouted order to reef a main topsail, and this during a gathering storm that now has the ship heeling hard over in a bitter cold south Atlantic gale, thrashing back and forth as the fierce swells shake with ceaseless violence the three thousand tons of iron-clad timber and her five thousand tons of Chile-bound cargo, and you do your best to climb as fast as you can up a rope ladder that is now stiffened by frost, the wind howling in your half-frozen ears, and then when you reach the yard to which your specified sail is clewed you must edge out sideways, on the port or starboard side as ordered by the bosun, until, as the vessel rolls, you find yourself hanging a hundred feet directly above the boiling sea, with just the yard you are clutching onto and a single ice-slick rope below you as a foot stay, and along with your brothers-in-arms beside you, you must reach down to the stiff and squall-wettened canvas sailcloth to where the reef lines are bent and haul together as one to bring the sail up to the yard and then secure it, tie it off so that the topsail is now shortened by the amount the skipper demands, to limit that acre of exposed sail by just a morsel so as to ease the strain and take account of the rising wind and the dangers that too much exposed canvas can pose to the stability of the vessel and her company. Only when that is all done may you climb back down below, except that if the wind strengthens still more and the captain wants another set of sails shortened still further and the bells have not yet sounded for the end of your watch, you have to go aloft yet again, risking, with your fatigue and your ice-cold hands and immovable fingers and frozen seaboots, falling off into the bottomless ocean—except that you won't and don't and you never do, and you are soon done with this task too and are back down below once again, and this time and at last eight bells have sounded and you, still in your soaking oilskins and your sea-filled seaboots, are in your bunk fast asleep while the ocean rages around you on the other side of that foot-thick hull and its iron cladding, keeping you and the *Peking* safe for another hundred miles as you edge ever closer to the promised calms of the Pacific beyond.

And then at last you achieve your goal; you are suddenly in this brand-new sapphire sea, and the captain orders a hard turn to starboard and the wind nearly immediately abates and the sea stills herself to a slow and steady swell, and the sun breaks through the scudding clouds and suddenly the air feels warm again. There are lights onshore twinkling a welcome and the windless world lies before you. Tugs appear, puffing gouts of steam; there is a dockside, some cranes, stevedores shout a casual Spanish greeting as if you had appeared from just across the river, and you tie up alongside and step gingerly down the unusually unshifting companionway onto the strange stability of solid earth once more, to a place of hotels and elevators and taxicabs and men in business suits, and you stand in a dockside bar and look back and note with approval and relief that the *Peking* is still there, her four masts towering above all other superstructures in the docks, yet quite empty of sails; she looks a little weary, a bit battered about, in need of a coat of paint or two, and quite probably the sailmakers below will be taking advantage of the sojourn here in Chile, for this is the Port of Valparaiso—the Valley of Paradise—where they can run up another half-acre of canvas to be quite sure the suite of sails for the return trip is as strong and secure and blazing white as befits a vessel as noble as the *Peking* of Hamburg, flagship of Germany's famous Flying P Line.

And yes, that makes you think, there is a return trip due, and soon. The cranes are already in their agnostic way plucking great metal-banded crates, items of German engineering this time no doubt, from deep within the holds, and outside glowing in the sunset beside the railroad track there are pyramids of hessian sacks, bulky with the fertilizer the farmers of Saxony and Hesse, noting on their calendars the coming northern planting season, are said to be yearning for. So maybe a week will pass, and you can take a few glasses of beer or a pisco sour and spend a while dancing with a pretty Chilean girl in a flowered frock and high heels before the call goes out for all hands once again, and you will be back aboard among the smells of rope and salt and Stockholm tar—and fresh marine-gauge paint—and hear the order to ease springs, and there will be farewells waved, the *Peking* will edge out into the roads, pass between the buoys, and, taking directions from the pilot and the harbormaster, will ease into the open Pacific Ocean, unfurl such sails as are necessary, and turn left, to the south and then east once again, bound for the Atlantic.

The seas will rise and there will be foam and spume and spray and swells and the soaring indifference of a following albatross or two, gazing across as you suffer your way through the passage. Only this time the land will be on your left, the Cape Horn light will be blinking more of a farewell than a warning, and, with welcome mercy for all, the wind will be at your back—at least until you turn through the strait and leave Patagonia and Argentina behind you. Your sails, all thirty-eight of them now fully rigged and none of them reefed, will send you flying homeward under the impress of a constant fair wind, as all sailors everywhere on the world's ocean would devoutly wish.

{ 16 }

The Age of Sail came to an end quite abruptly. The date on which a ship could, at least in theory, be propelled by forces other than wind is widely acknowledged: May 4, 1776. This was the date when, after many vexing years of failed attempts, James Watt perfected the steam engine. And the device that could then power factories and lift heavy things and propel vehicles could also be installed inside a ship's hull and made to turn a shaft, at the end of which would be a propellor, a screw, that would drive the vessel forward. No further need to fret about the trade winds or the westerlies or the doldrums; providing you had the fuel aboard—the coal at first, then, once it had been discovered, the oil of today—a powered ship could proceed anywhere, at any time, in any season, winds blowing or not.

But come the middle of the twentieth century, alarm bells began ringing for another reason altogether: the oily sludge, bunker fuel, that was being burned to power these marine engines that by then were crossing the oceans in the thousands was found to be causing immense amounts of pollution. And while at first that pollution was seen as merely cosmetically disagreeable, by the end of the century it was being shown to cause something far worse—the problem that dogs the world today and is a matter of urgent international anxiety: global warming. The fifty thousand mighty steamships that in the 2020s carry eleven billion tons of cargo each year—the ships getting ever more gigantic, those that carried a thousand containers in the 1960s supersized to load twenty thousand in 2010—were belching greenhouse gases in prodigious abundance. One large container ship is said to pollute as much each year as do fifty million cars, and all told, the world's shipping fleets—which also include bulk carriers and tankers—contribute around 3 percent of the planet's greenhouse gases.

A solution was sought. A new and sustainable kind of fuel might be burned—except that the kerosene-like substance known to be the cleanest is around five times as costly as normal bunker fuel, and shipping companies are reluctant. Then again, the various organizations that police the world's oceanic trade suggest they may enforce new regulations to lower emissions dramatically, but there is little optimism that such will be obeyed. Only one possible solution suggests itself: somehow employing the free, clean, and eternal force of wind once more and persuading it to help ships move along, just as they used to. Not by returning to the world of the square-rigger but somehow incorporating wind power into new designs for cargo ships that already use conventional fuel-burning engines to keep their propellors turning.

Having said that, a very small number of maritime idealists are, at the time of writing, going cold turkey and making trade arrangements that rely once again on sail and sail alone. Two French companies, Grain de Sail and TOWT, have built aluminum-hulled schooners—in the case of Grain de Sail, three of them, of ever-increasing size, starting with a seventy-five-foot-long boat capable of carrying fifty tons of cargo, and with the most recent model adapted to carry containers—albeit just two hundred of them, so no real competition with those behemoths that can carry a thousand times as many. The firm, which is based in Brittany and uses St. Malo as its home port, makes two or three sailing voyages a year, tracing a modern version of the infamous "triangle trade," but on this occasion exporting organic French wines to New York, taking American manufactured goods down to the French Caribbean possessions of Guadeloupe and Martinique, and then loading coffee and cacao beans for the three-week voyage back to France, where the company's founding twin brothers make organic coffee and chocolate for sale in gourmet stores in Paris and beyond.

The economically more rational and somewhat less precious reintroduction of wind to the international shipping business comes in the form of a still-developing technology known as WAPS, or wind-assisted propulsion systems. This involves retrofitting sailing devices of various designs onto the upper decks of existing vessels and using wind's helping hands to lower the use of bunker fuel; and, to the delight of the environmental community, significantly reducing the emission of greenhouse gases—with some sail designs lowering pollution by as much as a third. The wind-powering devices offered by the WAPS makers scarcely resemble the sails of old, looking like giant circular pillars bolted onto the decks or kites flying ahead of the ships with their positions maintained by lines that are controlled by computers on the bridge. Indeed, computers control every imaginable aspect of modern sail-assisted technologies—computers predict the winds en route, alter course, change the sails' trim and direction, and reef them as needed. No frozen seamen have to clamber up masts or risk life and limb on a heaving sea during a storm. The bosun will be sitting in the warmth of the bridge, coffee in hand, watching as the screens before him order the unfurling of a sail from a vast cylinder down on the deck a hundred yards ahead of him, and he will feel a moment of pleasure as the giant ship gives a nudge and a heel as the wind kicks in and the engines lower their growl just a little as free energy takes over and transports the great vessel for the coming hundreds of miles, if the wind is fair. And if it is not, then the touch of a button will furl it back into its container, the engineers will spool up once more, and the ship will be carving her way through the waves just as before, under steam.

{ 17 }

Time was when warfare was conducted under sail, of course—and a much more complicated relationship it was between the master of a ship sent out to fight and the weather in which he happened to do so. Cargo might be delivered later, or lost, because of an adverse wind; so battles could also be won or lost—wars as well—and the fate of entire countries could depend on the direction and the speed of the wind that blew over the sea at the time the guns began to fire.

My own experience of a sailing warship was, at least in the sense of the battlespace, vicarious at best. It was the occasion of the 236th anniversary of the Declaration of Independence, made by patriotic American separatists intent on wresting themselves free from the rule of the British Crown—July 4, 2012—when I, hitherto a London-born Briton, became officially transformed into an American citizen. I swore an oath formally abandoning any lingering fealty to any *foreign prince or potentate*, as Her now late Majesty Queen Elizabeth II was described, and formally threw in my lot with the young republic of the United States of America.

And I did so, my hand on my heart and repeating the solemn words of the oath spoken before a small crowd of like-minded new immigrants—a Vietnamese girl of fifteen the youngest, me by far the most geriatric—by a federal judge formally attired in her robes of office, on the afterdeck of the oldest surviving fully commissioned sailing warship in the world, the USS *Constitution*. It was blisteringly hot. The few sails above us moved listlessly as the ship rolled near-imperceptibly in the slight harbor swell. The crew, all of them formally dressed in eighteenth-century uniforms made of thick wool, were very evidently suffering from the heat, as were my guests—one of whom, Sir Brian Urquhart, was nonetheless so moved by the ceremony that he applied to follow suit and became an American citizen five years later, in his late nineties.

The *Constitution* is a three-masted, forty-four-gun heavy frigate, one of six ordered at the beginning of the nineteenth century by a republic just getting into her first foreign military engagement, against the Barbary pirates off the North African coast. Thomas Jefferson ordered his sailors into battle, and the *Constitution* played a part; her most notable achievement was during the 1812 war against the British, in which she overcame four Royal Naval vessels, most notably a slightly smaller thirty-eight-gun frigate that the British had earlier captured from the French, HMS *Guerriere*.

The engagement was, by naval battle standards, a fairly simple one, a classic mid-ocean encounter that pitted two roughly equivalent vessels against each other. It was fought on August 19, 1812, in the far western Atlantic Ocean some four hundred miles off the coast of Nova Scotia. The Royal Naval vessel was on her way to Halifax for a refit when first she spotted sails to her south, on her leeward side. The weather was cloudy but with good visibility, the seas had but a light chop, the wind was a fresh breeze, maybe Beaufort Force Four, from the north. The *Guerriere* slowed down and all looked off to port, when one of the sailors spotted a flag being raised: an American battle ensign. From her silhouette, this was evidently the *Constitution*—a frigate that *Guerriere* had herself been trying to catch some weeks before when she was part of the squadron sent to round up such American warships as lay lurking in this part of the ocean. There had been a chase, infuriatingly slow since the winds at the time were fitful and often calm, leading the American captain to lower the whaleboats and have sailors row her toward windier weather. The captain of the *Constitution* ordered his men to dump ten tons of her drinking water over the side, lightening her enough to allow her to outrun her British pursuer and escape into the safety of Boston Harbor.

So *Guerriere* was given permission to detach and proceed to Halifax and deal on her own with such problems as might befall her. Recall, incidentally, that in 1812 ships could communicate only when in sight of one another, by semaphore, by signal flags, by lights (though it would be another twenty years before Samuel Morse would create his code, until then light signaling was a primitive affair). And now here was a problem: *Constitution* was on the prowl, and she had half a dozen more guns than those at the British captain's disposal. There was a certain inevitability to the outcome. This would not be a chase, like last time. It would be a battle. A battle royal.

The two vessels closed, their gun crews rushing below to load their cannon with shot, the fire crews on each ship unrolling their hoses and lining up buckets of sand at the ready, the medics swarming to the surgery, for naval injuries tended to be non-trivial, with large wounds and broken bones. An American broadside could unleash more than 750 pounds of hot iron at the flanks of a British ship, and mortal wounds—to crew and, probably, to the weaker ship—were certain.

Which is, essentially, what happened. The American ship unleashed a volley of shot and chain shot at the rigging of the British vessel, and in short order brought down her mizzenmast. She then backed away, turned into the wind, and attacked again from the other flank, bringing down the foremast and the mainmast, rendering the *Guerriere* incapable of maneuvering—remembering that in 1812 it was wind and only wind that allowed a ship to move at any sensible speed. *Constitution* then came alongside, too close, and her bowsprit became entangled in the *Guerriere*'s dangling mizzenmast rigging. Marines—both ships carried a detachment—swarmed across this accidental bridge and there was hand-to-hand fighting, before a sudden exchange of fire from a revitalized British crew caused the American captain, after axing the bowsprit free, to pull away, head around the stern of the British ship, and pour musket fire into the officers' cabins at the stern. She then put about once again and was returning to the fight when she suddenly saw the *Guerriere* fire a single cannon shot into the sea away from the direction of the battle—which they took as a sign, if not a traditional one, that the fight was over, and the British would now strike their flag.

The Americans sent over a rowboat and boarded, asking that the British ship formally surrender, which her captain, with a sardonic remark, agreed to do. The two hundred British sailors (and by chance a small number of impressed Americans, whom the British had been gallant enough to keep from the fight so they would not have to engage with their fellow countrymen) were taken aboard, though the American captain refused to accept the defeated British officer's sword, since he believed he had fought with great gallantry.

There was an attempt to take *Guerriere* as a prize and tow her to Boston, but she was too badly damaged to survive the trip, and it was decided, after the Britons were allowed to retrieve their valuables and personal items, that the ship would be set on fire and scuttled. Her remaining ammunition exploded soon after the fires were lit, and the *Constitution*—damaged, but not excessively so—headed off southwestward and reached Boston ten days later, to scenes of wild enthusiasm. Her twenty-two-inch-thick oak hull had seen the British cannonballs bounce off like hail on a haybarn roof, prompting one wag to declare that her hull must have been made of iron. *Old Ironsides* is the nick-name she was given at the time, and the name survives still, proudly declared on her sailors' badges. She still sails when the winds are fair, showing herself off to the citizens of Boston every few years. At the time of writing her captain is, for the first time in her history, a woman.

{ 18 }

On a sailing warship there exists essentially only a single set of potential points of failure: the masts. Damage the rudder, and another can be jury-rigged with ease. Kill the captain, and the deputy is a trained replacement. Run out of shot, and you can easily turn tail and flee the scene. But if you lose the mast—or, as in the case of the *Guerriere*, all three of your masts—then the jig is up. All of a sudden you are quite unable to move. You are in an instant transformed from a nimble and mobile fighting machine into a sitting duck, an easy target. The three essential options for any warship, usually pinned to a bridge bulkhead, are that the vessel be able to fight, to move, and to float. If she is unable to move—because no masts means no sails, and no sails means no propulsion from the wind—then it is regrettably axiomatic that no matter how hard her crew may fight, she will soon and inevitably be so pummeled by her enemy that her ability to float will very smartly come to an end as well. To engage in a fight, you must have the wind at your beck and call; remove it from the equation and you cease, in essence, to be a warship at all.

The history of the Age of Sail is peppered with stories of

great sea battles in which the outcomes have been determined by the ways of the wind. In the case of this one fairly uncomplicated Anglo-American encounter, the outcome was ultimately laid at the door of the British captain's inability to make proper use of the wind. In other episodes, both before and since, it was an unexpected change in the wind's strength and direction that turned the tables between the fleets, or, in the case of what follows, determined the fate of an entire nation.

Kublai Khan, the ambitious reforming Yuan emperor of thirteenth-century China—as a Mongolian, the first foreign leader that China had ever had, and yet a man who brought a strange unity to a hitherto fractured nation—was persuaded by his courtiers that his potential for expansion, for foreign conquests, was in theory just about limitless. From visitors like Marco Polo, the West knew of him well; Coleridge would have him decree *a stately pleasure dome* be built in the expansion he called Xanadu. He already had possession of China and Korea; now, if he put his mind to it, he could be the absolute ruler of Vietnam, of Malaya, of Cambodia, and, most priceless of all, of Japan. And so, incautiously—history would later say foolishly—he bought into the idea. He assembled a fleet of no fewer than nine hundred Korean-built sailing ships, crammed them with fifty thousand soldiers from China and Mongolia, added another seven thousand seamen, and headed out from the ports in eastern Korea across the Sea of Japan toward the enormous Hakata Bay, at the head of which stands one of Japan's largest cities, Fukuoka.

The expedition went well enough at first: the Great Khan took with ease the two Japanese islands of Tsushima and Iki, and when the fleet arrived and landed en masse at the small fishing ports that dotted the circumference of Hakata Bay, the guardian samurai, unused to invasion on such a scale, fought as well as they could but were far outnumbered and rapidly pushed back into the countryside. It was November 19, 1274, and by nightfall it looked as though Kublai Khan might well get his way.

But out at sea the Korean pilots, good seamen all, thought otherwise. They sensed a change in the wind. An ominous change. They sent urgent messages to shore telling the Mongol armies to reboard the vessels and leave, for fear that the boats would be trapped on the lee shore on the western side of the bay. Some few hundred cavalrymen did make it to the ships, which all began, in haste, to leave the bay for the open sea. But it was too late. The storm was rapidly upon them, and it was then realized that the speedily built boats, ordered in too large a number for the Korean shipyards to make with sufficient care, were too flimsy to survive a massive windstorm; and so in their scores, and in the pitch-dark, they began to sink, dashing onto the rocks the seamen and the soldiers.

By morning the Japanese defenders emerged from sheltering in the woods and saw an extraordinary scene: hundreds of Mongol boats scattered or wrecked around the bay, and those few surviving craft still battling against the winds and trying desperately to make their way out to sea. The assault had been a complete and abject failure: Japan had survived the impertinence of an invasion by an alien power—and had been saved, in the view of the gleeful samurai, by the great good fortune of a change in the wind. A change brought about by the gods. It was a divine wind—a *kamikaze.*

The Mongols were a persistent people. Though the Great Khan had lost thirteen thousand men in his spectacularly failed assault on Hakata Bay, he decided to try his luck once more, seven years later. This time he would make doubly certain his fleets would leave well before the summer typhoon season, and in June 1281 he launched two separate armadas toward the coast of southern Kyushu. They made their first landfall at Imari Bay, fifty miles west of the previous disaster; here they were held off for no less than six weeks by a newly resolute samurai force, emboldened by the memory of their initial defeat in the first attack. The delay brought with it the prospect of inclement weather—and indeed, on August 16, a typhoon struck. The enormous invading fleet scattered much as its predecessor had done, and the ships were smashed once again on the rocks of the myriad reefs at the bay entrance. Thousands of soldiers were left to the tender mercies of the samurai—Kublai Khan himself managing to escape aboard an undamaged and still seaworthy boat.

The kamikaze had saved Japan once again, its divinity now firmly welded into the Japanese story. The word would later be used in another context during the closing months of the Second World War, the historical concept of the wind's fortuitous divinity employed to persuade civilians and airmen alike of the enduring nobility of the cause.

{ 19 }

Winds have played significant roles in more than a few naval battles elsewhere down the centuries. Long before the affairs in western Japan, Greece escaped what most Greeks today imagine would have been a fate much worse than death when the dreaded Persians threatened them on their very doorstep at the Battle of Salamis, in late September 480 BCE.

Ten years earlier, the Athenians had repelled a Persian attack at the Battle of Marathon. Now Xerxes, leader of the Achaemenid Empire, was back again, and with an enormous naval force of some 800 ships, compared with the paltry 378 mustered by Greece. He was bent on punishment, revenge, and subjugation that, had it succeeded, would have had world-shaking implications, considering that Greek philosophy and governance was at the time coming into full flower and would have been utterly negated had the Persians won the day.

But, in what is generally regarded as one of the most decisive naval battles of the ancient world, they didn't, and by reading the runes most naval historians today blame the sudden onset of fitful and contrary winds for scuppering the Persian battle plans. The literature of the time tells of much cunning in the opposing commanders' various tactics—secret messages, feints, servants pretending disloyalties among the various enemy participants—but the bottom line was simple: a combination of hubris (Xerxes had his imperial throne set up on a mountainside so that he could watch the battle from a place of landlubberly comfort), too many lee shores (the huge Persian fleet was tempted to muster in a narrow strait such that the ships were all pinned by dangerous rocks on both sides), and a clash between ships that were unwieldy (the Persian triremes were huge and cumbersome) and ships that were nimble (their Greek opposite numbers, though a much lesser force, were small and agile and easily maneuverable) led to a decisive Greek victory.

The action took place no more than ten miles from Athens, in the narrow strait to the west of the city that separates the mainland from the island of Salamis. The Persian triremes—two sails and some two hundred rowers stacked in three levels—thundered up from the south and squeezed through the strait into a small embayment. The Greek fleet then arranged itself to pin the Persians in this confined area, goading them into mounting an attack in the strait itself—at which point the winds changed, becoming fitful, and together with the visitors' evident unfamiliarity with a severe swell that runs through the strait a few hours after early autumn dawns, in a matter of moments the Persian fleet, their sails now flapping uselessly, suddenly lost its cohesion: their commanders ordering their men to row this way and that, the huge boats running into the rocks ashore, boats running into one another, boats turning turtle, the whole armada in disarray. Meanwhile, most of the Greeks watched with sardonic amusement, the wind evidently helping the Achaemenids destroy themselves with almost no Athenian input whatsoever. Mind you, one Greek captain did spot an opening and made a dash for a Persian ship and rammed it from the side, at one fell swoop smashing all its oars on that side and sending it careering to the bottom.

As best they could, the sorry remnants of the Persian fleet took to their heels and limped back to Turkey and, in time, to the coasts of Persia. Xerxes himself went overland for much of the way, and the Persians never again tried their hand at running Greece. The citizens of the Athenian city-state took on that task instead, building as quickly as they could a fleet that would manage the country's eventual vast maritime empire—which itself would help finance the beginnings of true Athenian democracy. Something that most probably never would have happened if King Xerxes had won the day in late September 480 BCE, but whose ambitions were vitiated by the sudden and historic change in the direction and tempo of the wind.

{ 20 }

And as the Greeks, then and still today, had ample cause to bless the wind, so too, a full millennium later, had the English. At the time—the late sultry summer of 1588—the country was terrified by the very real possibility of an invasion by fifty thousand Spanish soldiers, with an immense Spanish armada of some 122 warships heading up from Lisbon to defeat the British Navy and allow soldiers to be brought from the European mainland to land on English soil and kick-start the actual invasion. The purpose was simple enough: to dethrone Queen Elizabeth—a Protestant, like her father, King Henry VIII—and restore to England the role of the Roman Catholic church.

The violently bitter dispute over which branch of Christianity would rule England had been simmering for years. By 1588, after so many arrests and beheadings and burnings at the stake, all England knew that King Philip II would act decisively for the restoration of the local authority of Rome—whose pope,

Sixtus V, had promised a reward of a million ducats for the venture's success. An invasion fleet would inevitably be sent northward. The British forces waited for it off the southwest coast of England, bent on intercepting any Spanish warship that might appear on the southern horizon.

And waiting too was the wind. What would later be called the Protestant Wind, since what it did in support of Britain would help keep intact the power and authority of the Church of England, then as now the dominant religious authority in the land.

But before the wind got itself involved, there was battling to do. For days the British fleet waited in the Channel for a sign of enemy sails, with, famously, Francis Drake playing a nonchalant game of bowls on Plymouth Hoe while a gathering of nervous admirals fretted about him. Come July 19, the first of the square-rigged sails was spotted, looming northward off the Scilly Isles. The news was instantly reported by a fast messenger ship racing to the village of Kynance, where an emergency beacon was lighted, followed by a series of like-sized fires on the highest summits between Dartmoor and the Chiltern Hills, such that London knew within the hour. Drake himself, after casually finishing his game,* was taken to his galleon *Revenge*, and battles were joined.

There were three encounters. The first, off the port of Plymouth itself, was really more of a harassment by the Britons of the incoming Spanish fleet, with little damage done to either side. The encounter did, however, serve to warn the Spaniards against trying anything along the southern English coast and instead to make directly for France, where they were to embark the thousands of soldiers who, according to King Philip's battle plan, they would then ferry to the British east coast to begin their adventure. Drake and his superior at sea, Lord Howard of

---

* He lost.

While the defeat of the Spanish Armada in 1588, depicted here more than fancifully, still swells the hearts of all true Englishmen, it was contrary winds that wrought the most damage on the intruders and sent them limping home to Cádiz.

Effingham, however, had other plans. Once the Armada reached the supposed safety of the port of Calais, the British commanders sent fire boats—half a dozen still-just-floating wrecks that had been jam-packed with barrels of tar and gunpowder and all set ablaze—toward the anchored armada, causing them to go wild with panic, cut their anchor chains and cables, and sail hastily back out into the Channel.

But now the winds had strengthened and veered from their previous southwesterly direction and were now blowing more from the north, pressing the discombobulated Spanish crews and their boats against the dangerous chalk cliffs of what had swiftly become a lee shore. The English fleet was now behind them, enjoying the advantage of what is known as having the weather gage, being upwind and essentially in control of the battle. The fight, renowned in history as the Battle of Gravelines, a coastal village in what is now Belgium, ended all hope of even a modest Spanish success. They lost a significant number of their most valuable capital ships, and for one good reason: being downwind of the English, they had to twist and turn on the sea, and in heeling over, sometimes at considerable angles, the English gunners simply fired broadsides as low down as possible onto the Spaniards' exposed flanks such that when on the next tack they righted themselves, the holes would be below the waterline, with catastrophic results.

It was time for them to go home—except that the winds had other plans. For suddenly the breezes stiffened still further and backed toward the west, then to the previous southwest, and finally to the south. There was for the visitors no ready way home—the wind was contrary, the English lay in wait, and, if that was not enough, time at sea was taking its toll on the armada, with some of the vessels' timbers opening and warping and some entire hulls having to be sheaved together with cables. The reduced and disheartened remains of the fleet now had no choice but to head northward on their southerly wind, along the English coast to their left—past Norfolk and Suffolk, Lincolnshire and Yorkshire, Durham and Northumberland, with following English ships watching like sheepdogs to make sure no attempt was made to land any of the Flanders soldiers they had embarked at Calais before the fire ships put an end to their loading.

Queen Elizabeth, still not wholly convinced that the visitors had left, then came down to Deptford on the Thames and made perhaps the most famous speech of her reign, declaring to an adoring crowd the lines that would be quoted down the centuries by the sensitive everywhere: *I know I have the body of a weak and feeble woman; but I have the heart and stomach of a king—and of a King of England too, and think foul scorn that Parma or Spain, or any prince of Europe should dare to invade the borders of my realm.*

She need not have worried. The winds were pushing the would-be invaders steadily and relentlessly away. The threat wholly evaporated in mid-August when the armada passed abeam of Edinburgh, confirming their presence in the coastal waters of Scotland, then independent of England and, being Catholic, notionally friendly. The English fleet then broke away; the defeated

Spaniards, seeking only to get their ragged remains back home to the Galician port of A Coruña, headed up past Aberdeen and Inverness, crossed without harm through the notoriously violent waters of the Pentland Firth, then turned left and so headed south, passing the Hebrides, past Kintyre, beating down the west coast of Ireland.

It was here that the wind decided to give them a real and lethal walloping. Not the wind alone, however; not only did a series of formidable storms blow up, with westerlies pounding on the members of the scattered fleet and doing their best to drive the ships onto the knife-sharp gneisses of the Irish coast, but the mariners commanding the depleted fleet were puzzled and frightened by a strange and inexplicable *submarine* force that seemed to be trying to drive their vessels both backward up north and eastward, toward the rocks. This, though they didn't know it, was the yet-to-be-discovered Gulf Stream, well on its invisible way between the Caribbean and the Orkney Islands. The combination of this current and the westerlies took the most terrible toll on the hapless fleet. Ship after ship was torn to shreds—twenty-eight of them by most counts, far more than had ever been sunk at the hands of Howard and Drake. Five thousand men were drowned in raging waters that had been whipped up by an unusually ferocious chain of storms, noted by other log-keeping sailors in other sea lanes nearby.

The costliest of the wrecks was that of the fifty-gun galleass* *La Girona*. She had already been badly damaged, her rudder broken by one enormous Atlantic wave, and her captain had put her into a sea loch near Londonderry for repairs. Assuming—

---

* This was a hybrid kind of warship, initially operated by the Venetians, with lots of sails above for speed, augmented by a small number of oarsmen below to enable, in theory, swift changes in direction. The idea was not a consummate success and never truly caught on—the wreck of *La Girona* hardly enhancing the design's reputation.

this now being late October—that there was worse weather still to come, he opted to run back northward instead, heading for calmer waters and the benevolence of the Scots. He never made it. Powerful winds drove him and his fully laden ship onto what are now called the Spanish Rocks, by the tiny town of Ballintoy in Northern Ireland's County Antrim. Of the 1,300 men aboard, only 9 survived; the bodies of most of the rest were never found, although some 260 were buried in a mass grave in a local Irish churchyard.

It took some while for news of the calamity to reach Spain, and to bring a dose of cold reality to the court of King Philip. As he waited, nervously, he remarked that "I hope God has not permitted so much evil" as the rumors were beginning to swirl. Then, one by one, through September and October, his ravaged ships started to cross the harbor defenses and regain their country, coming not into Santander and A Coruña in the Spanish north but into Lisbon, from where most had set sail the previous May. Of the 122 ships that had entered the English Channel and were sighted on July 29, a scant 66—according to the most recent research; the figures are much debated—returned to Spain. Many of those were beyond repair and sank, rotted, or were scuttled. And so far as human casualties were concerned, the current figures hold that 25,696 men left the Spanish ports in May, and 13,399 returned. The loss of so many men and of a full third of the armada left King Philip in an inconsolable state, and he locked himself away in the monastery within the newly built complex of El Escorial to reflect on the loss. When he emerged, he laid no blame on his own navigators, mariners, and fighters; nor did he scourge his enemies, who had waged their battles with honor. There was but one party upon whom the responsibility for the failure fell. It was almost a heresy to say so, especially for one so devout as King Philip, but he said it anyway, in a single poignant line that reverberates down the years:

I sent the Armada against men, not against God's winds. . . .

{ 21 }

Four centuries later, with the Age of Sail well and truly over, wind was a much-reduced factor in deciding the outcome of battles at sea. Yet it did still have an effect, and in a somewhat unexpected way it played a crucial role in the greatest military expedition in modern history, the D-Day invasion of Nazi-occupied Europe in June 1944.

The statistics still make for remarkable reading. On the day of the landing itself, 156,000 troops* from the US, the UK, France, and Canada took part. They were brought to the five famous beaches—codenamed Omaha, Utah, Juno, Sword, and Gold—in five thousand ships. Eleven thousand aircraft crossed from dozens of secret British airfields. Twenty-four thousand paratroopers were dropped behind enemy lines. All of this was accomplished—under the direction of US general Dwight Eisenhower—between about midnight and lunchtime on the appointed day. Over the coming month, once the beachhead had been fully established and secured, more than one and a half million men were landed, together with 150,000 vehicles and half a million tons of supplies. It was the turning point in a world war that had already lasted nearly four years. And matters truly did turn as a result: Paris was liberated in August 1944, Adolf Hitler committed suicide in late April 1945, Berlin fell to the Allies on May 2, 1945, and Germany surrendered, ending the European war for good five days later.

---

* One was my father, a very young second lieutenant who went by ship with the 1st Battalion the Bedfordshire & Hertfordshire Regiment. He had been training for the invasion since January at Catterick Garrison in Yorkshire, one of 1,100 such training bases across the country. He went over on June 6, was captured on June 9 (D plus three), and was then spirited away to a POW camp in eastern Germany, to be liberated by American forces in May 1945. He was quite unaware that I had been born the previous September, and we met for the first time when I was eight months old. He never spoke about his wartime experiences.

And all of this might well not have happened at all—and most certainly would not have happened on this particular schedule—but for the work of a then-young Austrian-born oceanographer named Walter Munk. He would help solve—or at least obviate—the problem that my father and thousands of other frightened infantrymen had long fretted about: how would the troops manage to disembark from the landing craft if the winds were too strong, the waves were too high, and the sea was too deep?

Walter Munk—who wore with ease the sobriquet of *Einstein of the Oceans*, would spend his exceptionally long life—he died in 2019 aged 101—studying waves and swell and the winds that formed them both. For most of his career he researched at the Scripps Institution of Oceanography in La Jolla, California, outside San Diego, working on long-distance submarine sound communication and the generation and passage of waves. He was especially interested in how waves move across the ocean—and to follow and measure the passage of a train of waves he set up six well-equipped monitoring stations across the entire Pacific Ocean, from New Zealand to Alaska by way of Samoa and Hawaii, that tracked the changing wave heights generated by a violent storm in the Southern Ocean, off Antarctica. He differentiated between ocean swells, which are low-frequency emanations from distant storms, and waves, which are shorter and steeper and may almost vanish at the surface but retain their integrity for surprisingly long distances. He divided the marine real estate into sea, shore, and surf, and became able to predict with uncanny precision the arrival time of waves of various heights and frequencies in each of those zones—thus becoming, inadvertently, the darling of the surfing communities around the world, with youngsters who would regularly tap into his predictions to find out where and when the best surfing waves might develop. He was seldom wrong.

His work on wave prediction began during the war, when Munk left his civilian academic studies and joined the US Army—unusual among the Scripps community, most of whom not unnaturally preferred the navy. The sea, however, pulled him back, and he was temporarily assigned to the US Navy's Radio and Sound Laboratory, where he and his PhD supervisor Harald Sverdrup became fascinated by the idea of being able to forecast wave heights. This became especially relevant in early 1944 when he heard through the grapevine that the Allies might be planning to land a vast armada of men and matériel by sea somewhere in Nazi-occupied Europe. Troops had already made landings on the North African shores with some success—though the Mediterranean is a much more benign ocean than the Atlantic, largely because the *fetch* of the waves—the distance they have to travel from their point of origin to where they are observed—is necessarily much smaller. Munk was convinced they could help any Allied effort in France, and he sent messages offering his help to the Operation Overlord planners in London. Until Sverdrup, then the director of Scripps, became officially involved, Munk—an Austrian émigré, no less, and so on paper hardly the most credible character—was rebuffed. Once his academic chops were acknowledged and accepted, the importance of his work became suddenly apparent.

For tens of thousands of men and machines were about to be shipped aboard hundreds and hundreds of hastily built flat-bottomed landing craft, vessels that had been designed to carry them from their anchored destroyers and frigates and battleships and corvettes and minesweepers and troop carriers that had brought them over the sea and then slide them toward the heavily defended beaches of Normandy. As soon as the landing craft steersman believed the waters beneath his hull were shallow enough, he would order the huge steel bow gate to be dropped open and the men and their vehicles would be ordered to move ahead without delay and rush en masse down the ramp into which the gate had now been transformed and into the green and opaque mystery of the seashore water that extended for some way before them—a few feet if the steersman was prudent, a daunting many yards if not. And the craft would not run themselves aground to make the soldiers' lives easy, since each had to reverse and go back to the ship for more human cargo, making return trip after return trip to collect yet more men and yet more machines to fill the beaches with still more fighting power, and it would do no good if the vessel was stuck fast on the beach with the tide falling beneath its hull and so likely to remain stranded high and dry and vulnerable, sitting-duck style, to being obliterated by German cannon fire.

The sharp-end-of-voyage details involved here: How deep might the waters be? Could the infantrymen even swim if they had to? What would happen to a tank if a huge wave dropped thousands of gallons of water down its turret? Might the average soldier, knocked off his feet by a rogue swell, be swept away by a coastal current and landed, spluttering and angry like a wet hen, dozens of yards away from where he was meant to be? Had such matters been considered by the planners of the forthcoming battle, or had they been discounted, or forgotten, or ignored?

The generals and admirals and senior staff officers back in London would speak airily of armies and corps and divisions and brigades and battalions, would refer to infantry and artillery and engineers and signals and catering—but did not initially think too deeply of what it might actually *be like* for a young soldier, like my frightened and apprehensive father, though he never spoke of his fears, laden with weapons and ammunition and rations and trenching tools and wearing a heavy wool uniform and stiff leather and iron-shod boots, to leave the safety of the end of the ramp and find himself in ten feet of cold water, gasping for air, panicking despite all the drills, not knowing if he could or would ever surface because the waves would knock him backward and make him stumble and crawl underwater on his knees and wrap him in a miasma of sand and seaweed and the bodies of colleagues who had just been shot and lay floating or sinking all around him.

Walter Munk, five thousand miles away in California, knew he could forecast what the waters would be like, depending on the observable and forecastable winds, and so could help the planners avoid such nightmares for the men who would be at the cutting edge, the very vanguard of the invasion. At first his offer of help was politely rebuffed; once his supervisor wrote to support him, however, the Overlord staff capitulated and agreed to accept the American's advice. From the early summer of 1944 Munk's input was sought by the British meteorologists advising the planners, for they accepted that his work meant he knew what they could not: how the waves would be at the time and date that had been chosen for the invasion.

Initially, General Eisenhower chose Monday, June 5. The movement orders were issued, the planes were fueled, the soldiers embarked on their ships, the gliders attached to their tows, the paratroopers readied for their nighttime drops, the French resistance agents were alerted (by the broadcast of the second line of a Verlaine poem), the sabotage raids set in train, the hospitals readied for mass casualties.

The weekend weather was poor; it rained heavily in London, and the Channel was windy and choppy. Moreover, three depressions had formed near Baffin Island in Canada and were approaching the British Isles at speed. It looked as the though the operation would have to canceled—a monstrous inconvenience that would have had a savage effect on the morale of the waiting troops and would risk the security of an operation that now, because of the timing of the moon and the tides, would have to be delayed by at least two further weeks.

All eyes turned to Walter Munk. His advice, filtered through the British weather planners, was stark: the waves would be too fierce for the landing craft on Monday. But during Monday night and into Tuesday morning there would be a break, a window of opportunity when the winds would abate and turn in such a way that the waves off the Normandy coast would be much reduced.

The period of relative calm would not endure for long, but if the generals and admirals could stand the strain on their patience, Tuesday, June 6—a delay of just twenty-four hours—was, in his considered view, prudent and advisable.

Standing before all the grandees of the operation, Eisenhower read and reread the cables and signals and heard out the views of the fliers and the sailors and the weathermen. And then he decided: Operation Overlord would be stalled by one single day. It would now go ahead on Tuesday, June 6. He issued the formal order. The paratroopers, headed for positions behind enemy lines, would thus have to leave in a matter of minutes. All the others could grab a few more hours of quiet. But now, and at last, the operation was on. The greatest military invasion in modern world history was formally set into terrible motion.

And Walter Munk would, from this moment on, be forever known as the Man Who Delayed D-Day.

{ 22 }

Other invasions were more insidious. The world has suffered over the centuries with countless colonies, and the siting of all too many of them were consequent upon invaders arriving there by sailing ship, propelled by the patterns of the wind. The Age of Sail and the Age of Exploration—during which so much settling and annexation and expropriation and colonization took place—were essentially coterminous. Any imperial possession that had a coastline, for example, had first to be reached by alien sailing ships that had been blown there—and mostly by happenstance—under the press of the wind. The foreigners arrived, found the place to their liking, and essentially said to themselves and to anyone within earshot *I'll have that*, ran up a flag, stationed an armed piquet, and seized it for themselves.

It would be valueless to linger too long on the matter of how often and how greedily Britain did precisely that. Though critics and admirers will bicker endlessly over this most incontinent of modern empires, the simple facts are undeniable: From Cape Town to Hong Kong, from Jamaica to Ceylon, from Australia to Newfoundland, from Guyana to Zanzibar, British sailing ships came across coastlines and harbors and evidence there of items and services they might well one day need—water, gold, indigo, rubber, ivory, sandalwood, opals and amethysts, and human beings who might repair their ships or service their sailors' carnal needs, or else might be put in chains and taken off to be bought and sold as movable property. Of British imperial perfidy too much has been written, and too often; and besides, the case of Portugal is simpler and more illustrative of the manner in which windblown sailing ships took their imperial urges across oceans such that all Portuguese today admit to the lament: *So small a country to live in, but yet the whole world to die in.*

Portuguese sailors had been trading in England since the thirteenth century, selling dried fruit from the Algarve, salt from Setúbal, a red dye called *kermes*, and lengths of leather from hardy mountain cattle. They also had a maritime insurance company, and over the coming decades were in the process of creating a fleet of vessels specifically designed to explore the possibility of further trade with merchants of North Africa. They had already heard rumors of great treasure in eastern and southern Africa, but had for years been unable to reach there by ship because of one notorious maritime obstacle: those places were beset by particularly extreme and fickle winds.

The original Arabic name for Cape Bojador is *Abu Khatar*, which translates to the Father of Danger. On a map it appears utterly insignificant—merely a modest pimple on the great bulge of the Sahara that pushes west into the Atlantic. It doesn't look at all impressive when spotted from the deck of a sailing ship either. Such a vessel, an early version of the fast caravels the Portuguese were now building in large numbers, would be dipping happily along, sailing on a southwesterly tack parallel to the Moroccan coast against the prevailing westerlies at maybe eight knots, the sky blue, the desert dunes off to port rising reddish orange, the sea green where it was shallow and deep blue well off to starboard. Then the navigator would spot the cape and sound a warning: sand dunes ahead, beaches, a scattering of jet-black rocks topped by scrub, then a line of breakers stretching westward out to sea. At first blush this was certainly no Cape Horn, though its significance to mariners had for many years before been quite as mythic. *Nothing to see here, just move on,* one might be tempted to mutter.

Except that four things then happened in quick succession. A white smoke seemed to start rising in great gouts from the surface of the water ahead. A strange hissing, punctuated by curious slapping sounds, began to reverberate, clear as a bell, from up ahead. The ship was promptly caught in the grip of a southward current that seemed to draw it toward these mysterious and nameless perils off the bow. And, most significant and inexplicable, the wind direction suddenly changed, shifting from the west, veering through the compass until it was blowing lustily from the northeast, filling the sails and pushing the little ship out into the open ocean, drawing her away from the relative security of the shore.

Back then, mariners seldom liked to move out of a comfort zone that involved the sight of land—a phenomenon known to many even to this day, where cruise-line passengers confess to feeling a deep unease when first they cannot spy land on any horizon. Today it is easy to calm such fears by mentioning radar and radio and GPS and lifeboats and safety-at-sea protocols; back in the fourteenth century, though, the absence of land was for a seafaring innocent a truly terrifying thing. And to compound it here, with white smoke and hissing waters and mysterious currents—small wonder sailors were afraid of whatever was going on at Cape Bojador and opted by the score to turn tail and flee back up to the safety of home and assume, as most did, that beyond the smoke and the hissing lay an ocean full of sea monsters and lethal, ship-swallowing dangers.

And then came a sailor named Gil Eanes. It was 1434, and he was a junior functionary in the service of Henry the Navigator. Having been humiliated by one failed attempt to round the cape the year before, he approached it for a second time with grim determination—finding it to be more feared than fearsome. The white smoke was simply spray and Saharan-sand-laden spume blowing up from a submerged reef that extended invisibly for some half a dozen miles off the coast; the sounds of hissing and flapping were caused—incredibly—by millions of *sardines* that liked to frolic and feed in the upwelling clouds of nutrient-rich waters here, with the said sardines in turn being chased and mostly eaten by carnivorous predators who followed them through the foam; the current was found to be a northbound offshoot of another equatorial waterway, in due course, once the oceanographers got to work, and thus easily explicable; and the northeasterly winds were the soon-to-be-famous trade winds—which all, other than those still spooked by their sudden appearance, blessed as offering them a new and fair-wind trade route across the Atlantic, leading (in the specific case of the Portuguese) to the treasures and tragedies of the vast land they would come to call Brazil.

All Eanes had to do was to accept the folly in his fears, head west where the newfound winds were blowing him, grit his teeth, and remain steadfast on the now landless open ocean for a day or so until the waters to his south had cleared of smoke and hissing and underwater reefs, then turn his rudder and tack until he was heading south once more, and then southeast—until the coast of almost-sub-Saharan Africa appeared before him once again. He and his crew would scan the land for any signs of human habitation from which he might learn or else subdue, would look onshore for some token to hand to Dom Henry, and then head home around a cape that now held no fears for him or for any Portuguese mariner or explorer to come. He eventually presented his sponsor with a garland of wild African roses, and at the same time gave Portugal the keys to all of the windswept world.

And they took it with both hands. For the five and half following centuries—until as recently as 1999, when Lisbon returned to China the tiny coastal enclave of Macau—Portugal swept the seas for useful places to exploit, inhabit, or own. Tentatively at first, then with increasing confidence, the country's sailors advanced southward down the African coast by about one latitude degree a year, building up forts and trading posts as they went—Madeira, the Azores, the Cape Verde Islands, Angola, then around the Cape of Good Hope to Mozambique and, once they had worked out the advantages of waiting for the southwesterlies that brought the monsoon to India, tore their caravels across the Indian Ocean to build a series of ports and enclaves from Goa to Cochin, Daman, and Diu, around the coast to Malacca and Timor and all the way to Nagasaki in Japan. Whether denoting full-scale possessions or factories or simple trading houses, the flag of Portugal rose in abundance just about everywhere the winds had blown their ships ashore. Other empires—the Spanish, of course, the Dutch, the French, and, inevitably, the British—were rivals, and over the years the fate of Lisbon's possessions waxed and waned with the accidents of history. But of all the world's great empires, there can be no gainsaying: Portugal's was the one that most decidedly, during its accretion and consolidation, had the wind firmly and profitably at its back.

{ 23 }

Here we now leave the sea and return to the land. Across which, wind being wind, the wind still blows.

Come to the Texas Hill Country, or to the high plains of Nebraska, or to where, on a scorching day, cattle stand under the shade of old oak trees or shelterbelt cottonwoods ringing the pastures of Oklahoma or Kansas. In all such places and a thousand others besides, dotted around the American prairies, you will still see all the old and lasting emblems of pioneering times, as though Willa Cather and John Steinbeck and Sinclair Lewis had never passed on, were writing columns for the local papers telling of hardscrabble life as it still is, unchanged, eternal. The tumbleweed still bounces across the landscape, the dust devils scoot up small tornadoes on the die-straight dirt roads, there are longhorn cattle skulls and glinting lines of barbed wire and drying pools with flocks of wading birds pecking away at the mud.

And then there is the sound—two sounds, really, a combination of one that is both dependent on and coexistent with the other. There is the wind, soughing and whistling overhead, blowing for days on end, hot and dry and dusty. And then there is, on almost every farm and ranch and smallholding between the Rocky Mountains and the Mississippi River, the whirring, clacking, clattering, creaking, groaning, gushing sound of an American prairie wind pump, a water pump powered by wind.

There is something infinitely comforting about the sight of such a machine—usually a tall steel frame, or maybe an older version made of trestle wood, and at the top a ten-foot-diameter wheel with eighteen—it seems invariably to be eighteen—curved steel sails or blades; and behind them a glinting steel box, sometimes silver-colored, sometimes red, that conceals the gearing that you know somehow causes the steel rod that is visible within the tower to move up and down, up and down, a couple of feet each time, lifting water from the bottom of the well up and out through a pipe that gushes it, cool and fresh and clear, into an aluminum holding tank around which a small congregation of cattle or horses or maybe sheep is gathered, and who may drink from it from time to time as the mood takes them. And comforting? It is because a working water pump on a Texas farm signals the presence of life in a place that in ordinary circumstances would be hard-pressed to sustain it. Such pumps made it possible for the prairies to be settled in the first place and fully and efficiently farmed—with highly disagreeable consequences for the native people who had lived there before white settlement, as all must now agree—but a symbol nonetheless, for good or ill, of progress.

One name dominates the field. Up at the top of the tower, standing directly behind the steel case that houses the mysterious gearing mechanism, is an eight-foot-long metal vane that keeps the sails facing into the wind, no matter how swiftly or erratically it might change direction. Stenciled in enormous letters on both sides of the vane is the word that is as emblematic of the windmill as the windmill is itself: AERMOTOR. Once there were hundreds of small companies making wind pumps across the pioneer states of America; now there are only a handful, with the Aermotor Windmill Company of San Angelo, Texas,

The Aermotor Windmill Company, now of San Angelo in West Texas, has been making all-metal wind pumps for prairie farms since 1888. The clacking sound of the blades turning in a low breeze stirs immediate memories for all dry-land farmers everywhere.

the best-known of the survivors. This small firm has been making such machines since 1888. Many of those that still rattle and squeal and whirr while hoisting water from deep in the parched prairie earth are themselves well over a century old. Just as the Texas farmer likes to think of himself, and just like the Texas wind, these rickety-looking devices just never give up.

{ 24 }

While Thor Heyerdahl may well be right in his previously noted claim that *man hoisted sail long before he saddled a horse*, the harnessing of wind by those who live on the land, far from any water—in fact, *needing* water, as many rural Texans do—has been going on for centuries. Although historians like to point to places now called Persia and Afghanistan (but, rather counterintuitively, not China or Mesopotamia) as locations where wind-powered machines first

The flat and low countryside of the Netherlands has for centuries demanded the use of large mills to grind the grain and raise the water from the polders. Perhaps nowhere else on Earth has such a long historic association with the power of wind.

found favor, northern Europe seems to be where the boldest advances have been made.

Nowhere is the wind and the machines that employ its power more firmly associated with the national character than in what is called the Netherlands. Not for nothing does the Dutch word translate as the Low Countries—since a formidable amount of the national territory is close to or below sea level, and machines powered by the wind that blows constantly overhead have been used for centuries to pump the water out to sea and, protected by high dikes and dams, allow the land to be reclaimed and dry out. The *polders*, as the dried-out lands are known, are a beloved essential of the country's very existence: no matter how different the ideological views of the country's politicians may be, all are wholly agreed on the existential necessity of the polders and what more broadly is known as the polder system. The central role that wind has played in creating the Dutch polders and, until modern technology took over, sustaining and protecting them is never challenged. The Netherlands is a nation eternally threatened by water but invariably saved by wind.

And not only saved but fed as well. Items of Dutch-made wind-powered equipment have been used for hundreds of years not just to deal with water—to pump it either onto the land, as in Texas, or away from the land, as in Holland—but to perform a task known by one of the oldest words in the English and Dutch languages: to *mill*. Milling grain to make flour has long been a special task for which wind has been employed in those European places where it blows regularly with sufficient force to be made use of. The Netherlands, being flat and exposed to the near-constant westerly gales, has long been an ideal site for that uniquely and emblematically Dutch machine, the windmill.*

They can still be seen all around the countryside, piercing

---

* To the pedant, the Aermotor Windmill Company does not, in fact, make windmills but wind pumps.

the horizon like fortresses. Their design has scarcely changed in centuries. The Rijksmuseum in Amsterdam has dozens of windmill images in oils from the seventeenth century and the so-called Golden Age of Dutch painting—and in each the design looks little changed from those structures still visible today. The main body is a tapering cylinder of stone, like an enormous old pepper pot, maybe fifty feet wide at its base and eighty feet tall. It has a rounded cap at the top, and, halfway down the body, a wooden gallery that entirely surrounds the mill, so that the miller can walk completely around his structure. The gallery is supported by struts that extend from its outer edges back to the mill's base, keeping it secure and helping with the stability of the tower itself. The tower is exceptionally sturdy, its stone exterior furnished around foundations of huge oak beams and posts, with thick pinewood floors, its gigantic inner gearwheels made of well-seasoned elm and its sails made of light and flexible ash.

There are invariably four sails extending from the cap of a Dutch windmill, with each sail a good forty feet long, its end meeting the level of the gallery so that the miller may inspect it close up when the mill is at rest and it is safe to do so, as it is each dawn when the mill begins its working day and the silence of a damp Dutch polder is broken by the distinctive swooshing and clicking sound of an old mill, in perfect working order, bending to the task of making flour, for which it had been designed three centuries before.

The sun will just be rising as the miller unlocks the two main wooden doors at the base of the tower, and against which some overeager farmers have left bags of freshly harvested durum wheat, hoping to steal a march on their neighbor-competitors and have it turned into flour before the day is done. The miller, a congenial fellow, harrumphs, knocks out his pipe on the door frame, and heaves the sacks onto the loading platform for later. He climbs the ladder to the gallery, opens its doors, and steps out onto the platform, the sun now fully risen and illuminating the village that sprawls down in the valley below and beyond him. Whoever built this mill set it up on a bluff, maybe sixty feet above the tributary of the Rhine that meanders through the flatlands here—the bluff being about as impressive an elevation as can be found in the billiard-table flatlands of this part of the country.

He looks up at the telltale by the top of the mill's cap and sees it streaming happily toward the southeast—indicating that the wind this morning is blowing from the northwest, maybe twenty degrees away from where it was blowing the day before. He now has to unlock the cap and shift the sails around by twenty degrees or so, such that the wind—if it stays in more or less the same direction all day—will blow directly at the four sails. A capstan and chain are set into the gallery to allow him to turn the cap—a heavy thing, cumbersome to move if the wind is blowing in an uncooperative direction. So he struggles a little, relieved that the gallery is not slippery with frost or snow, until finally he has the sails in place. He then locks and chains the cap in position, and then, moving them by hand one by one, brings the sails down to where he is standing so that he can unfurl the sailcloth and make the sails ready to perform their allotted work. The night before he had furled the cloths to one side of each sail, and the sails' ashwood framework then lay naked and open like a forty-foot-long ladder, transparent to the air so that any wind blowing onto it would gust right through, with no resistant sail surface to interrupt—and thus derive energy from—the onrushing gale.

It takes him five minutes to unfurl each length of sailcloth, and so about half an hour after he came into his mill the sails, now all facing directly into the wind like a ship's square-rigged sails, present four great surfaces ready to begin their turning. Already they are flexing and bowing outward under the wind's pressure, the halters and harnesses holding them back thrumming with energy and just waiting to start turning. Inside the mill all is now creaking, groaning, the ropes all tight, the gear wheels snug and ready to be unlocked, the grain bags ready to be hooked and chained up to the platform where the millstones will do their grinding to reduce the grain to kitchen-grade size and softness.

But nothing is going to happen until the miller has released the brake; and that he won't do until he has greased all the bearings with lard and wax, depending on which surface needs smoothing, and until he has made certain all is ready for the very considerable wind power to be safely and fully unleashed. Maybe one of the local millwrights, from the family that has for generations tended this venerable structure, has come to repair and tune up and maintain this particular mill, having been asked to stop by to make some small adjustment—a peg loose here, a broken gear there, or else someone is needed to slip into what the owner himself finds a tricky place to lubricate. Everything has to be in good order, *shipshape and Bristol fashion*, as they say, before the main brake is slipped and the sails slowly begin to turn, then steadily accelerate until, with a deafening swishing roar that indicates optimum speed, the day's milling gets underway.

For when a full gale is blowing outside, the inside of a running mill can be a daunting place. For a start, every piece within is just so massive. The nonmoving components are balks of hewn oak maybe forty or fifty feet long. Whirling elmwood gearwheels can be ten feet in diameter, their pegs or teeth six inches tall and all too easy to get caught up in. The sheer mass of the pieces within a mill is measured in tons, clearances in crawl spaces can often be minuscule, the noise is terrific, and the air is often full of wheat dust and chaff, making it in places near impossible to see.

The main brake itself, a long slither of cloth and leather pressed tight by a great oak lever, is kept chained and locked hard up against the sails' flywheel. Now, the checks all complete, it is slowly released, and one by one the sails start to swish down by the tiny window beside the storage bin, into which the grain from the overnight sacks has by now been tipped. A sail passes by every ten seconds, making one complete revolution every forty seconds—the wind, in other words, is brisk and robust, all the miller could hope for. The flywheel turns—counterclockwise seems to be the preferred direction—the enormous main shaft turns like the prop shaft in a powered ship, and then, by way of a series of elmwood gear trains, the turning force of the sails from their massive horizontal spindle is transmitted to another wheel that turns vertically; and this in turn has a second set of gear-wheels that mesh with a smaller, still vertical shaft that drives the millstone itself.

There are two millstones, each about six feet in diameter. The lower of the pair is fixed, locked into the mill's thick pine-wood floor. The upper stone, with spiral striations carved into its inward-facing surface, rotates, turned by the main shaft. Once the miller sees the rotation is smooth and regular, he opens the vibrating grain shoe, dispensing a stream of the freshly harvested wheat into the top aperture, from where it is spread in an instant between the two grinding wheels. With equal speed the spiral grooves on the upper millstone move and crush the wheat ever finer as the curved ridges and valleys move it from the stone's center to its outer edges. The ground meal then spills forcefully from between the stones' edges and into a second wooden catching chute, which directs the flour into bags. If all goes well, the miller, watching the flow of the meal with a keen eye, will fill and then switch to a new bag every two or so minutes, and with each one weighing forty pounds, will have produced two hundred pounds of whole wheat flour in no time at all. Perhaps the farmer who left the bags outside wants just white flour—in which case the mill owner will pour the whole wheat flour into the horizontal drum of a tumble separator, which is also run off leather straps connected to the main sails' rotation, and within moments he rebags it into white flour, semolina, and bran.

And thus does a windmill's operation continue, hour upon hour. Once in a while the wind may shift and the owner will have to use his crank wheel and chain to turn the cap and the sails through a few degrees to catch the full force again; but in places like Holland, mostly flat and with a generally equable climate devoid of drama—no hurricanes, few tornadoes—he rarely has to reset the mill entirely but rather works with the ebb and flow of a sustained current of cool air, turning the sails faster or more slowly depending on its speed and strength.

The village mill was once a hallowed place, an essential, a landmark, universally respected, to children a thing of awe. Bearing in mind that a mill was invariably confined to the countryside, it was often by far the tallest structure around, particularly if mounted on a hill or a bluff. The local church might have a taller tower or spire, pointing toward heaven, but it was usually built down below so that Sabbath parishioners might have easy access. The mill, however, enjoyed the solitude of altitude. Its isolation, its unique identity, caught the attention of more than a few artists—Rembrandt, of course, and his compatriot Jacob van Ruisdael and then later on Vincent van Gogh, who lovingly painted the great mill at the top of Montmartre, back when this hill was a place of gardens and allotments, with the only taller structure being the Basilica of Sacré Coeur. In England the windmill was an ever-present fascination of the pastoralist and romantic painters like Constable, J. M. W. Turner, and even Thomas Gainsborough. All such artists seemed to accept the slow-moving sails as somehow organic, an integral moving part of nature's panorama, and not necessarily as an industrial barbarism that might have attracted an artist like Gustave Doré. A seventeenth-century windmill was a comforting sight, a benison for any rural horizon, not too different from a sturdy and imperturbable oak or the vision of an elm tree swaying in a storm.

The Ingenious Gentleman Don Quixote of La Mancha may have thought of windmills somewhat differently. Cervantes, in his delicious early-seventeenth-century tragicomic fantasy, has the poor and deluded knight tilt his lance and charge at mills, which he saw as fantastical and menacing giants out to do him no good. His faithful servant Sancho Panza at his side, Quixote rode at full gallop upon his broken old carthorse, Rosinante, visiting no harm on the imperturbable mills but damaging his own dignity and amour propre.

The symbolism of the tale, which is still taught in some of the better schoolrooms of today, has long endured, however, along with the word *quixotic*. To take impetuous action, to do on a whim something built on a fantastical version of chivalry, to *tilt at windmills*, is to rage pointlessly at the inevitable, to pit idealism against reality in a vainglorious struggle that you, like Quixote, are sure to lose.

But there is another explanation offered for what might have been going on in Miguel Cervantes's mind: that Quixote was actually doing battle with *progress*, fighting against what in 1605, when *Don Quixote* was first published, was seen by some as a dangerous new technology. This might seem risible today; such classical windmills appear nowadays to be magnificently old-fashioned, and likely never to have been thought of by anyone as technologically sophisticated. To people who knew of sailing ships—and most in Britain and the Netherlands at the time would have been passingly familiar with the sight of great square-riggers being readied in port or else heeling before the wind offshore—a mill was in essence not too difficult a device to imagine: a logical development in the harnessing of the wind. Could there have ever been a time when people hated windmills, feared them, thought of them not as structures of elegance and beauty but as gigantic fortresses freighted with threat and menace? Maybe Cervantes was telling us something that we now find difficult to accept.

{ 25 }

If so, then the parallels with today are quite clear. For the windmill of old has been replaced on the modern skyline with a bright and shiny new object of immeasurable size that, though perhaps not too technologically astonishing—it still has sails turned by the wind—is very much emblematic of the times—and of times that, by comparison with those of seventeenth century Europe, are less restful, less certain, more dominated by public anxiety.

Today's windmills do not manufacture flour. They manufacture power.

A modern windmill—or wind turbine, to be exact—is not so much a construction that invites affection or radiates pastoral comfort. Rather, it is something built out of an urgent necessity—a need for a better means of generating electricity, an invention made to wean society away from polluting ourselves into oblivion, and as such is a device that triggers in the public mind a certain degree of apprehension, being a stern reminder of how we had all better shape up, or else. If Cervantes was right and seventeenth-century Spaniards did think of such mills as icons of menace, then some of us feel similarly today; except that the stakes—our very existence—are considerably higher.

Menacing to behold, maybe, but fast becoming near ubiquitous. At the time of writing there are said to be 341,000 such wind-powered turbines operating around the world, with 70,000 of them in the United States. Denmark generates more than half of its electricity from these behemoths. Scotland produces more electricity from wind than it knows what to do with, and so exports it to its neighbors, England most notably. Further statistics, rather too ephemeral to be included in a history, are perhaps better left to journalism. Suffice it to say that these new machines, now so huge as to resemble triffids or Martian invaders, infest both our landscapes and, more recently, our seascapes, and have a way—such is their unimaginable vastness—of dominating both. Wind farms, their makers like to call these infestations. Wind factories, which is what they more properly are, is a term of art seldom used.

It is more than a little surprising that the wind-powered

electricity generator took so long to invent. After all, as long ago as 1831 Michael Faraday had created the generator itself—producing measurable and then useful amounts of electricity by moving a magnet back and forth through a coil of copper wire. But then it turned out that converting the kinetic energy of movement into electrical current was not that easy to perfect; it took until 1870 before William Armstrong did it with a water-powered turbine, and three further years for Charles Parsons to achieve much the same thing with steam.

It was 1887, more than fifty years after Faraday, that a Scotsman named James Blyth used the kinetic energy of moving air—the wind blowing around his holiday cottage in northeastern Scotland—to generate electrical power for himself. His home-made machine managed to produce enough electricity to keep all ten of his incandescent light bulbs burning, and to power a small lathe—for no running cost whatsoever. The wind in

In 1887 the visionary Scottish engineer James Blyth built a wind-powered electricity generator above his summer cottage in the village of Marykirk—producing more than enough power for his own house and the rest of his street. The other villagers declined his offer; and so widespread was the skepticism that it was not until the 1960s that wind power took off.

Marykirk, Aberdeenshire, like the wind everywhere else in the world, was, at least ostensibly, the precious gift of nature, given away for nothing.

The context of this small piece of history has an inescapable irony to it. By the time of James Blyth, the Industrial Revolution was some ninety years old,* and going full tilt. The newfangled idea of generating electricity, which it was accepted would massively enhance industry's various processes, and generating it by the employment of steam-powered turbines, happened to find particular favor in the Scotland of the day, for one very good reason: Scotland was thick with coal. Steam was then best generated by water being boiled by the burning of coal, with which the Scottish lowlands in particular were abundantly supplied.

The result of this was that the newly formed fossil-fuel industry, in the person of the wealthy owners of Scottish coal mines, was an immediate and very keen lobbyist for steam-fueled power generation. One mine-owning Scottish duke had in one of his three castles south of Edinburgh a full-sized bath fashioned from a single piece of coal. To such people the idea of electricity being generated from something—like wind—that was free was an assault on the noble principles of capitalism, and a gross impertinence to boot. Just like today, when the big oil companies continue to do their level best to challenge those who support sustainable energy sources—wind being one, of course—so the battle was also being joined back in Victorian times. And so far as Professor James Blyth was concerned, the fossil-fuel lobby which on hearing of his invention decided to confront him, went on, at least in the short term, to claim victory.

* The favored birth date is May 4, 1776, when James Watt finally perfected his steam engine, with the critical help of John "Iron-Mad" Wilkinson, who showed him how to make a leakproof iron cylinder—after which sales of the Boulton & Watt engine went through the roof and mass production of devices made from interchangeable parts went into full swing.

{ 26 }

James Blyth taught engineering at a local Glasgow college and knew what to do to convert his fascination with the potential of wind power into reality. To the probable chagrin of his neighbors in the coastal village of Marykirk—where the winds from the North Sea are regular (and can be strikingly violent)—he decided in July 1887 to construct in the front garden of his little cottage on the main street a rickety wooden tower, more than thirty feet tall, looming over his roof. To this he attached four thirteen-foot-tall canvas sails that he suspended from steel arms, or whips, as he termed them. As the wind blew, so the sails turned, just as in a flour mill, rotating as they did so a hefty metal spindle that, through a series of gears, then turned a vertical spindle; at its base, by way of yet more gears, the rotation was converted back into the movement of a second horizontal rod that rotated a massive iron flywheel. This in turn was linked by a stout rope to a so-called Burgin dynamo, a state-of-the-art direct-current electrical generator with coils of copper wire turning between the wings of a powerful magnet—and which, true to Faraday's predictions all those years before, produced in a pair of stout copper wires that constant stream of flowing electrons (flowing one way—hence DC rather than AC) that we now know as *electricity*.

Blyth was a canny fellow. Excited though he may have been to now have a source of power for his little cottage, he did not immediately connect the wires that carried it to his light bulbs or the power tools in his workshop. To do that would have limited his nighttime illumination to those moments when the wind outside was blowing and the canvas sails were turning. Such would be a major inconvenience if, let us say, he was reading, deep into the best part of a good book, and because outside all unexpectedly fell calm, the sails above his roof slowed and stopped, causing the lights inside to promptly dim and then go out altogether. This problem he solved—clever Scots engineer that he was—by hooking up his dynamo to a cluster of imported and newly invented French *accumulators*, the forerunners of modern rechargeable batteries. This arrangement meant he could make use of his electrical power as and when he needed it. The accumulators would keep his electricity stored for many days, maybe even weeks, no matter whether the wind up above was blowing or not.

Moreover, his windmill and dynamo worked so well, the North Sea winds were so robust and his own domestic requirements so modest, that he found himself in the happy position of having electricity to spare, and could offer it around to his neighbors. But—and here the meddling of the fossil-fuel industry has to be a suspect—someone had put around word that electricity so made was *the devil's work*. He offered to wire up, connect, and illuminate all the streetlights along Marykirk's village center, but the town fathers, under pressure from an unspecified quarter, declined his offer, and the roadway remained dark and sunk in sodden gloom for the entire quarter century during which Professor Blyth's wee house was lit and cozily inviting.

He was not one to care unduly. He patented his invention of a *wind engine*—his patent number GB19401 of November 1891 representing therefore the *fons et origo* of today's vastly important wind energy industry. He went on to build a very much larger version of his domestic wind generator for the one organization that would accept his largesse—the Montrose Lunatic Asylum. His design for this machine differed from his first, having eight half-barrel-shaped sails mounted on a vertical axis—much like the Thomas Robinson anemometer, to be described in a later chapter. Being mounted on a vertical spindle, it did not need to be turned into the wind, as traditional windmills do and anemometers do not, and it ran happily for twenty-seven years, charging a battery of accumulators that brought light to the hospital's patients and staff until it was dismantled in 1914, eight years after its inventor's death "from apoplexy" in Glasgow in 1906.

{ 27 }

James Blyth's two generators were never equipped with brakes, and so it was feared that in a full gale one or the other might well go rogue and wreak great damage to buildings and passersby. As it happens, Blyth was lucky, and nothing of the sort ever occurred. But because a few months after Blyth's first generator was raised in Marykirk an American inventor named Charles Brush built a similar model in Ohio that did have a braking mechanism, he is often credited as the inventor of wind energy.

The Brush wind-driven generator supplied power to his enormous mansion in central Cleveland—he was already a successful and wealthy industrialist, his company producing arc lights and small steam-powered generating stations in New York and San Francisco. It was huge—twice as large as Blyth's and in many ways automatic, designed to face into the wind thanks to a rudder-like device behind the sails. And it worked perfectly, illuminating the ballrooms and dining rooms for the many guests who paid court to him at his Euclid Avenue estate. But he never took out a patent; and though he seems not to have suffered unduly from the opposition of the fossil firms the way Blyth did—no claims that his power was in any way diabolical—he has vanished from the list of pioneers who are so honored today, now that wind has become a fashionable essential. He eventually sold his company to Thomas Edison, helping thereby to found General Electric, and died in 1929, a formidably rich man*—but not the godfather of wind power. That was the rather more modestly situated Professor Blyth of Scotland.

---

* But also slightly mad—having first come up with a theory of gravitation that involved streams of tiny corpuscles hitting objects and subjecting them to forces that made them move in mysterious ways; and second claiming the existence of a new and very light gas to which he gave a name but that turned out to be a form of steam.

Which is where, sixty years later, the world's first commercial wind generating station was built. The Costa Head Wind Turbine was painstakingly constructed, under very trying conditions, in 1951 at the top of a cliff at the northern tip of the largest of the Orkney Islands in the far north of Scotland. It is a lonely and brutal place of mud and heather and vertical sandstone cliffs under endless assault by enormous Atlantic waves. But this otherwise wholly unappetizing place of seabirds and spray had then, and still has today, a feature of great potential value: it is just about the windiest place in the United Kingdom. The usual jokes are still made about the locals walking with a distinctive forward lean, a counter to their near-eternal gales.

The order for the mill was placed during a particularly bleak postwar period of economic austerity. Not a few of the local Orkney sheep farmers already had very small home-built wind generators to help with eking out their budgets for the costs of lighting their properties. This, combined with the fact that the main local electricity-generating utility—the government-owned North of Scotland Hydro-Electric Board—already generated most of its power from dams and waterfalls, and so was sympathetic to sustainably sourced generation (this was many decades before the discovery of the North Sea oil and gas fields, which sent Scotland firmly back into the arms of the fossil-fuel lobby), led to some outside-the-box thinking. Why not build a large experimental wind generator and connect it directly to the national grid? Such a thing had never been tried. But James Blyth, sixty years before, had shown that a generator could work at scale—so why not do it again, this time even bigger, bolder, and with benefits not to a single consumer in his holiday house but for the community as a whole and at large? No talk of *the devil's work* this time around: this being 1951, all knew that the devil had lately been in Germany and that he had been seen off. The war had cost money; the country sorely needed to be restored and rebuilt. Electricity was badly needed to help in the process, to kick-start Scotland's postwar economy—and it should be made cheaply and sustainably. Wind could possibly be the ideal way of making it.

The firm that won the contract was at the time the pride of Scottish engineering: John Brown and Company of Glasgow, whose shipbuilding arm had constructed some of the world's most famous ships—*Lusitania*, *Aquitania*, *Queen Mary*, *Queen Elizabeth*, and the Royal Navy's largest battleship, HMS *Hood*. On Costa Head the construction was very much more modest: a seventy-eight-foot tall, seven-ton tower of steel girders topped by a nacelle holding all the gearing and the governor and the generator itself, and then up front a three-bladed propellor, with each of the blades twenty-five feet long and made of laminated and reinforced wood. There was a brick building beside it with a variety of switches and controls that would allow the newly generated electricity to be fed into the main grid at the required transmission voltage and, crucially, with it being alternating current at the required frequency of 50 Hz, the flow of electrons turning back and forth fifty times a second.

Scratchy black-and-white films show the progress of the tower, the erection of which was kept largely secret from a British public that seldom visited the Orkney Islands, nor had much interest in going to so wretched a place as Costa Head. The working conditions—most particularly in winter and during a vicious snowstorm—would give an OSHA inspector apoplexy. Men in flat caps and smoking pipes would carry buckets of nuts and bolts up rickety-looking wooden ladders to fix the joints on scaffolding fifty feet high. Another man would straddle the smooth and slippery outside of the nacelle with studied nonchalance and with no thought of safety ropes or harnesses, holding on for dear life to any protruding spar each time a westerly snow squall carried his hat whirling down to the sea. Each finely shaped propellor blade was the height of four men; to see one John Brown worker clambering along it, similarly unprotected, with the hard and unforgiving headland dozens of feet below, and then a gust setting the whole arrangement moving—well, one just had to look away.

But it worked, and for a while at least, pretty well. A thirty-knot wind—a modest breeze, "a good drying day" for the housewives of Costa Head—would turn the propellor at 150 rpm, and step-up gears inside the nacelle would increase the generating speed to 750 rpm—more than sufficient to produce a constant output of 100 kilowatts. Moreover, through the attachment of a series of nifty new devices, the apparatus was made fully automatic: it would start itself whenever the wind blew favorably; it would stop itself if the blasts became dangerously unstable; it would point its blades in the optimal direction; it used a system of gears and levers to keep the propellor properly aligned with the airstream. Or, as a 1951 copy of *The Engineer* had it in what seems to have been something of a scoop, "inside each arm are fixed and moving pistons for varying the pitch of the blades. This is done hydraulically under the control of a servo-governor, the motion being transmitted through torque tubes and Hardy-Spicer couplings to quadrant boxes situated at the ends of the hub arms."

Yet it broke down more and more frequently as the months and the gales went on, and after two years of running—and more and more visits by exasperated technicians who did not find Costa Head the most congenial of places for a call-out—John Brown decided to pull the plug, literally. Pipe-smoking men in cloth caps came back and took everything down, leaving only one sad little brick building as memorial to the application of a technology whose time, in 1953, still had not come. That same year the nation looked to the coronation of a new queen and the climbing of Mount Everest as notable achievements and for a while forgot the events up in the Orkneys, which, if not exactly marked by failure, were not a stellar success either. Moreover, this turned out to be—coincidentally—the beginning of the end for John Brown, which was losing more and more shipbuilding contracts to yards in Korea and Japan, eventually closing up shop and shutting all its docks and graving yards along the River

Clyde and declaring bankruptcy. The engineering arm, pioneers to a greater degree than they ever knew, was sold to a Danish company and also ceased to exist.

{ 28 }

And then came October 1973, when President Nixon opted to support Israel during the brief Yom Kippur War with Egypt and Syria. The Arab oil producers had a fit and decided to punish the United States for doing such a thing by shutting off all petroleum exports. The resulting "oil shock" triggered a memorable autumn of what Americans saw as intolerable hardship. Gasoline, which for the better part of the previous fifty years had been cheap and plentiful, was suddenly nowhere to be found. I was living in a suburb of Washington, DC, at the time, and remember having to get up at 4:00 a.m. on those days when, because my license plate ended in an even number, I could go wait in line at my local Exxon station for a fill-up that set me back almost a dollar a gallon, when the going rate before the war had been thirty-two cents. Moreover, President Nixon ordered the national speed limit—even on the fancy new multilane roads of the Interstate Highway System—to be cut back to 55 mph, which seemed to most motorists like crawling through molasses. For three miserable months no one went anywhere—and a realization slowly and steadily dawned, based on an urgent new question: Was the reliance on oil such a good thing as had so long been advertised? Not only did the burning of gasoline cause pollution and ill health and maybe, as was being rumored, some possible unspecified alterations to the world's climate, but now a group in faraway Arabia, people of whom we knew little and yet had great disdain for, could stop our access to it at a moment's notice, making us feel vulnerable in a way that hitherto comfortably situated Americans were unaccustomed to.

Maybe some other source of energy, some alternative, based on a substance that was readily and inexpensively available, and one that it was not possible to ration or intercept or politicize or weaponize or use in such a way as to make us feel vulnerable—might it not be possible to find such a source of energy and see if it might be possible to employ it instead?

And then someone remembered Costa Head, and a whole new industry was born. That someone was a former US Navy officer and scion of a Wisconsin dairy farming family turned engineering professor at the University of Massachusetts Amherst named William Heronemus, who if not the father of wind power—James Blyth most assuredly deserves that somewhat clichéd distinction—is the man who dragged windmills into the space age.

Under the direction of Bill Heronemus, the image and efficiency of such wind-driven machines was profoundly changed, and almost overnight. Gone were the cumbersome and fragile-looking trestles and gantries of the Buster Keaton era. Gone were metal

Bill Heronemus, unchallenged as the founding father of the modern wind turbine industry, was a Wisconsin farmer, a US Navy destroyer sailor in the South Pacific, and an MIT engineering graduate. His adoring students called him simply "The Captain."

struts and canvas and laminated wood. Out was the cumbersome; in was the graceful. A tall and slender hundred-foot white pole formed the tower of the first such model; there were three blades, gleaming white and optically smooth, made of fiberglass or carbon fiber or some other new, strong, and very light material. The generating portion of the device was neat and tidy and housed in the drum-shaped hub where the blades came together. The machine looked all of a piece; it looked modern, serene, and, to be candid, *cool.* Its blades turned slowly and steadily and with a kind of lofty dignity, and it was controlled by invisible computers that kept it from over-speeding or overheating. Students at Amherst, where "Captain" Heronemus taught and designed with fantastic energy—huge multibladed floating wind generators for use on the sea surface was one idea that went nowhere—looked up at that first device (called WF-1, or by the students, Woof One) with awe and pride as it steadily beat out its characteristic thundering whisper. They could see the future, and it was tall and smooth and gleaming white.

In truth, this was not exactly the case with America's very first wind farm, which was built nearby in 1980 with some help from Heronemus's former students. It was set up on the two-thousand-foot summit of a ridge in south central New Hampshire, and consisted of twenty turbines mounted on an array of sixty-foot trestle towers, and it did not look modern, sleek, or elegant in the least—instead it rather resembled the work of Rube Goldberg or Heath Robinson. None of the turbines seemed to work properly; blades kept breaking and falling into the forest, and most of the time fewer than half of the generators seemed to be working, mainly because for some reason the average wind speed on the mountain was a mere 13 mph, and no one could explain why it had been chosen in the first place. The farm was a dud, in short, and its inexpressibly ugly towers were torn down for scrap after less than two years, having contributed precious little to the electrical grid, or to the landscape.

Their brief life did, however, mark the beginning of the wind power revolution, which continues apace to this day, seemingly unstoppably. Starting in the early 1980s, planners around the world—Denmark, Sweden, France, and Britain in Europe—working for huge private firms chose the sites for their new farms judiciously, based on a new generation of carefully crafted maps showing which places in the world were subject to the most reliable and strongest regular winds.

These firms created—in America, doing so with help from technically adept bodies like NASA—turbines that could gather wind and create power with infinitely greater efficiency than before. Firms with unfamiliar names like NextEra Energy and Terra-Gen Power built tremendously tall structures with impossibly large blades—up to four hundred feet long in some cases. They built on land, on high plains, and in mountain passes through which gales would howl incessantly—one such project being built in New Mexico by a company owned by Canada's main pension fund will have 950 generators, most of which will be put together by General Electric, in revived and dusted-off factories in southern Texas and upper New York State. And they are built not just on land but in the ocean, in the shallower coastal waters offshore—the wind being steadier and more reliable when it blows over water, less interrupted by hills and buildings. Some complain that their view of the oceanic horizon is spoiled by a distant line of poles with their blades slowly turning in friendly disagreement; some worry about birds being hit and frightened away; and when one blade broke south of Martha's Vineyard and scattered flotsam on the island beaches, some worried about pollution, and the government obligingly shut the generator field down for a few days, then realized the loss of all the clean energy that was being made, judged the risk acceptable, and allowed all the rotors to begin turning again.

On the dockside of Boston Harbor there is now an enormous brick building inside which turbine blades, some well over 180 feet long and—despite being hollow and made with three-inch-

Despite pressure from politicians, offshore wind farms have a certain inevitability about them, their backers insist, and images like this, of a sizable wind farm off the British coast, will soon become commonplace.

thick walls of a fiberglass, balsa wood, and Styrofoam sandwich—weighing twenty tons or more, are brought by ship to be tested in simulations of the most extreme wind forces they will likely ever encounter. For a million dollars and change, the engineers here will bring a blade inside their cavernous state-financed testing center and attach the blade's root—the big end, where it is normally joined to its brother pair to make up the wind-gathering trinity that stands at the very top of most turbines—to a thick cement wall, employing well over a hundred foot-long bolts to do so. Then up to three hundred sensors and strain gauges are fixed to critical places along the blade's length, after which an elaborate system of winches and cables and hydraulic motors are unleashed on the blade, which is held down like a floating Gulliver, with scores of minions who then scuttle and chatter around it to see how the giant beast will behave when finally it wakes up.

Which it does at the touch of a button high up in the computer-jammed control room, the operators all behind inches of armored ceramic glass, as the winches begin to turn and the hydraulic rams to oscillate and the cables to tauten—and the blade starts to come to life, to bend and stretch and flap and distort alarmingly as the worst imaginable winds start to buffet and batter it with unendurable force. At times the very tip of the blade—which, when it operates, will often reach speeds of 200 mph, even though seeming to rotate slowly—will be forced upward until it grazes the very roof fifty feet above and then thrashes back down to ground level like an enormous whip being cracked and cracked, for days and nights at a time. A test—the center can do three at a time—can last for many months.

The object is not to try the blade's patience and tolerance to destruction but to measure how it flexes and stretches and where its most vulnerable joints and junctions may be. But very occasionally a blade, inadequately made or designed, will suddenly explode under the enormous pressures—with a sound so colossal that the security guards at the dockyard's entrance call the police, afraid they heard a bomb go off, and all the dogs of Charlestown start barking with fright.

But the Boston WTTC, the only such testing center in the United States and one of only seven in the world, is now too small and no longer fit for purpose. Blades are getting so huge—one new prototype Chinese turbine blade made by Dongfeng Electric and destined for an offshore farm in the South China Sea is 420 feet long; and while the Chinse have their own test facilities for seeing how their blades behave, *thank you very much*, other manufacturers will need to ship their blades to America, so long as Boston has the capacity to do the work. Accordingly, a new building is underway; by the time this book appears in paperback it should be up and running, its blades flapping and flexing at scales never supposed possible even five years before.

At the time of writing, more than a tenth of America's power is generated by wind. Some European countries enjoy a far greater percentage—in blustery Denmark, half of the country's electricity is made by wind, in Germany a quarter, and down in Brazil and over in India 10 percent and growing. And in China, the rise of wind power generation and the size of the machines that perform the task is simply astronomical: from the western mountains of Yunnan and Sichuan—and in troubled Chinese-occupied Tibet, with its endlessly gale-swept plateau—to the coasts off Shanghai in the east and Hainan Island in the south, turbines of ever-increasing size are making their languid presence a wholly normal and acceptable feature of every skyline, every horizon.

Questions arise, of course. Birds fly into the blades and die; the noise emanating from the towers irritates some; others complain of the ruined view of the countryside; and oil leaking from the generators can discolor the blades and give the towers a look of decay. Some wonder how these immense blades will be dismantled and disposed of once their twenty-year life span is over. And there are the inevitable accidents: the components that need maintaining are at a great height, and workers who fall from the towers seldom survive. Few who saw the 2013 image of two young men hugging each other on the summit platform of a Dutch turbine as fire raged around the blades, trapping them, will ever forget it.

In all other senses, though, wind turbines are seen today as a consummate success—leaving only one question. Given that Michael Faraday invented the electrical generator in 1831, and that James Blyth illuminated his cottage in a Scottish village with a wind-powered generator just fifty years later, why did it take the world a further century for the potential of wind to be realized? The planet has suffered hugely during that one century, as vast amounts of fossil-fuel by-products have been created by the burning of all the coal and oil that long fueled the world's tens of thousands of power stations—when with some thought and imagination, the wind—free, clean, and blowing endlessly above us—could have been employed instead. It was an opportunity missed, with incalculable effects for us all, and for the tiny blue-and-green sphere that we like to call home. Countries have now changed policy, altered priorities, and changed direction. Let us hope we are still in time, and that it is not too late.

Meanwhile, as the gales grew stronger and more fierce, and as the authorities decided it best to let the public know of advancing winds so they might take appropriate action—launch sailboats or take cover, tie down roofs or bring out the lawn chairs—so efforts were made to assign numerical values to the different powers and strengths of the wind. Measurement of wind became a public fascination, and in some places remains so to this day, a surprisingly enduring aspect of the wind's enduring story.

CHAPTER FOUR

The Poetry of the Approximate

Nature, rightly questioned, never lies.

—Robert Mallet, *A Manual of Scientific Enquiry* (London, 1849), et seq.

{1}

The following few pages are intended mainly to celebrate the life and admire the achievements of Rear Admiral Sir Francis Beaufort, whose association with the classification of the winds is familiar to many well beyond the communities of the scientific and the weather-bound.

What his admirers most cherish about his main legacy—his deservedly world-famed Beaufort scale—is its determined imprecision, the artful vagueness with which this consummate seaman described the way in which air and water behave while the stream of moving wind increases its force, its speed, and its power. This imprecision, admirers are known to feel, offers a sense of the ineffable but enduring joy—the word seems fair to use—behind the admiral and the work for which he is so revered.

Context demands, though, that we look first at the lives of a smattering of other figures—and one in particular—to whom precision, in the specific matter of the measurement of wind, was by contrast quite essential. Though overlooked and all but forgotten, this particular "*large man with a natural animism*," according to the *Dictionary of National Biography*, was Thomas Romney Robinson, and he invented the anemometer. And just like Francis Beaufort, he was an Irishman.

{ 2 }

The Northern Irish city of Armagh—Queen Elizabeth II declared it to have earned formal city status in 1994—has in its two thousand years of history suffered a powerful lot of religion and amassed a goodly dose of science. Though fewer than twenty thousand souls inhabit Armagh, it sports two cathedrals, both of them named for the St. Patrick who is Ireland's patron saint. One, the older and less flamboyantly attractive, is the Protestant version; the newer, with

Thomas Romney Robinson, from the Irish cathedral city of Armagh, the center of both religion and science, was obsessed with finding a means of accurately measuring the velocity of invisible wind; the Robinson anemometer, still an essential tool for the serious meteorologist, is his best-known legacy.

two needle-tipped spires, promotes the Roman Catholic faith—and has the larger congregation, since 70 percent of Armagh's population subscribes to the ecclesiastic authority of Rome. The archbishops are each titled Primate of All Ireland, and in the pecking order of both churches are senior to their opposite numbers in Dublin, Ireland's capital a hundred miles and a world away to the south.

To those of a non-churchly leaning, such an arrangement appears unnecessarily unwieldy. Where scriptural dogma used to meet science in the city of Armagh, the unwieldy verged on what we today might think the insane. The great Archbishop James Ussher, who held the All Ireland Primateship starting in 1625, is best-known for his sedulously scripted calculation that he insisted proved—insanely—the date and moment of the Earth's creation. The blessed event occurred, he declared, at eventide on October 22, 4004 BCE. For many years this date was inscribed in rubric red in most Bibles around the world, and such was the good archbishop's authority within the church that for years the only challenge to his assertion was whether God had created our planetary home—with all its wind and weather—before His dinner on the twenty-second or whether he had done so shortly after He consumed His breakfast that same morning. After which, of course, came the six days of frantic creative activity, followed by a well-deserved rest on the seventh.

A century later, real science was to purge such thinking from the city and clerisy of Armagh, as in many places elsewhere, during the ferment of the Enlightenment. Particularly so in 1789, when an exceedingly Enlightened archbishop, Richard Robinson (who was no relation to our inventor), created an institution, the Armagh Observatory, which still stands today as one of the United Kingdom's hardest-working centers of astronomy. Archbishop Robinson, who was not Irish but came from the great Rokeby Hall estate in north Yorkshire, and from which he would take his later title as the first Lord Rokeby, suffered from a kind of architectural incontinence, building smart and useful structures all over town and turning a hitherto rather ordinary Irish community into as handsome a place as befitted the spiritual center of Ireland. He built, at his own expense, a public library, a county jail, a hospital, a school, and a bishop's palace, and then threw up a hundred-foot-tall marble obelisk for no better reason than to commemorate his friendship with a faraway English neighbor-aristocrat, the Duke of Northumberland.

It was in his observatory that Thomas Romney Robinson, the good bishop's unrelated namesake, eventually developed his celebrated invention.

The purpose of this device—which he called the *anemometer*, a word he unashamedly borrowed from the Italians, who four centuries beforehand had experimented, somewhat unsatisfactorily, with a variety of cumbersome and inelegant meteorological instruments—was quite simple: it was designed to measure, or to *meter*, the velocity of the movement of air that had been spurred on by the *anemoi*, the wind gods of ancient Greece. There had been efforts to do so before Robinson solved the problem in 1846. The best-known of the Italians was Leon Battista Alberti, a Renaissance architect who tried to signify wind speed using the same technique as a cobbler or a wig maker or a pawnbroker might unintentionally do, and whose suspended storefront symbols—a pair of boots,* a peruke, a trio of hollow brass spheres—would be blown by the wind such that passersby would have some apprehension of the weather just by glancing upward. Suspend some kind of thin metal sheet on a spindle, said Signor Alberti, and let it be blown against a scale so that its deflection between vertical (no wind) and horizontal (a full gale) would match the ferocity of

* As in the opening scene of David Lean's classic 1954 romance *Hobson's Choice*, which features a pair of high black boots swinging violently in a Salford rainstorm above Hobson's doomed shoemaker's shop.

the blast.* Ingenious, complicated, but somewhat limited in failing to offer up any kind of notion of the wind's actual velocity, other than that which was compared with itself. Yet it remained the preferred method of wind measurement for at least two centuries, from when the idea was first bruited in 1450 until the mid-seventeenth century, when the great polymathic astronomer and microscopist Robert Hooke employed an Alberti device when he needed to experiment on his landmark studies relating to the physical nature of air.

{ 3 }

Over the following inquiring, post-Enlightenment years there were other measuring devices and ideas, some of them elegant, many pedestrian. One of the more imaginative came from a canny Scottish amateur observer named Alexander Brice, who lived in the hills south of Edinburgh (high enough, he wrote, "to expose me to every blast that blows . . . the force of which I am, so often, obliged to feel"). He made in particular one most valuable observation—and then a series of most useful measurements.

Mr. Brice's observation, of a kind for which enthusiastic amateurs were then much valued,† held that it was pointless to try to

---

* The basic principle still extends to the classroom today: a Ping-Pong ball on a string will be deflected by the merest puff of wind, and its deflection can be measured on a scale.

† Science teachers at my boarding school, back in the 1950s, encouraged amateur observation. We devised one small scheme that involved collecting many thousands of used local bus tickets and adding together the six digits printed on each ticket, which ranged from 000001 to 999999, giving a range of totals from 1 to 54. Plotting these totals on a graph should have shown a peak at the mean of 27—but it didn't, the peak being skewed slightly lower, at around 24. We concluded that ticket sellers threw away the tightly rolled inner spools of tickets as they curled and proved too difficult to hand over—and these tickets, at the very end of each

obtain an accurate measurement of wind speed at ground level because the wind here was invariably capricious and fitful and given to wild variations in speed and direction as it bounced around off trees and hills and buildings big and small. But higher up above the steeples and the treetops the clouds seemed not to be affected at all by the caprices down below but sailed on serenely, white and puffily delicate against the raw blue of the sky, moving along in one direction only, uninterrupted by any deflective inconveniences. Moreover, depending on the angle of the sun, the shadows of these clouds moved across the fields and pastures and meadows of his modest Kirknewton mansion—and if only he could devise some kind of scale, then he could measure the time it took for the leading edge of one or more of these cloud shadows to pass between a pair of fixed and marked points, perform a simple mensuration and arrive at a speed of the wind in miles per hour.

Accordingly, in a pasture that he could see clearly from his dining room window, and beside which stood his impeccably accurate grandfather clock, he paced out and then marked with stones and sticks a straight line, to which he took a tape and measured it in feet and inches. It came in at exactly 1,384 feet, a little less than a quarter of a mile.

He then waited for a suitable day, and at the very end of March 1763 he was suitably rewarded: the day dawned bitter cold, the hills covered with snow, but with sunshine and just a few small clouds, all of them being blown eastward by a relentless full gale. Brice took station by the north-facing sash window and watched as the edge of one small cloud intersected with his line of stones, and using the tick of the clock standing to the right of the window he timed how long it took for the edge to move from one end of the line to the other. It was exactly fifteen seconds.

---

roll, had the higher numbers, all in the 990000 range. The bus company designed new ticket machines and sent us a pound to thank us for our pains.

He repeated the observation that day with ten other shadows, and each one of them took a quarter of a minute to pass over the same quarter-mile stretch of pasture—so, a quarter of a mile in a quarter of a minute. Very roughly, a mile a minute, or sixty miles an hour. In fact his calculations were rather more exact; and in his letter to Sir John Pringle, who at the time was president of the Royal Society, he recorded that the wind blew that day at 62.9 mph; and that on another day in May it blew at less than 10 mph. In his view, this was an impeccable means of measuring, with considerable accuracy, the velocity of the wind, requiring no equipment other than a field, some good boots, and a line of stones, as well a fine-running clock. And a mansion, of course, though of modest scale.

The Royal Society duly published Brice's communication—quite possibly because the author obligingly added a bonus, calculating, from his own observations once more, how much liquid rain was equivalent to a certain footage of snowfall (of which his Scottish hilltops received an abundance). The section in his paper on this matter remains a little confusing to the lay reader, but it seems to have worked out that ten inches of snow was a frozen and fluffed-up version of about one inch of rain. The manner in which Brice described his determination of wind speed—and to a single decimal place, 62.9 mph—is, by contrast, easy enough to apprehend.

{4}

But a still better way was sought, as anemometry evolved steadily into a minor eighteenth-century obsession. Some attempts verged on the wholly mad. Fit men carrying light cotton flags were invited to run *against* the direction of a hilltop wind until their flags drooped listlessly as the runners' speed exactly matched that of the wind into which they were running: know your own speed (by timing your run and knowing its length) and you could realize that of the wind. Somehow the scientific community became seized with the impracticality of this method, but it took time, so eager were they for a solution.

That a wind exerted pressure that will vary in line with its velocity—obvious to most who observe ships' sails and the vanes of a windmill bending to it, but through much of history a generally uncalibrated reality of physics—offered still further means of determining its speed. The French engineer Henri Pitot discovered the relationship between speed and pressure while working in Paris to measure the flow of the River Seine. The mercury in a diaphragm-sealed vacuum tube lowered into the river would rise according to the square of the speed of the water entering the tube's open mouth. Such a device, when much later installed out in the open air and attached to a wind wave so that its open end always pointed into the breeze, would react instantly to changes in the wind's speed. The mercury would rise up a carefully calibrated tube, and by reading the number and then performing a simple square root calculation, the wind's speed could be determined in a heartbeat. The sensitivity of the device was the key—every gust, every puff could be recorded almost as it happened, in real time—which is why a pair of pitot tubes, as the instruments are now known, are fixed to the outside of the nose of almost every serious aircraft ever made, displaying to the pilots in the cockpit how fast the wind is blowing against the fuselage, and thus how fast the aircraft is flying compared to the air through which it is doing so.

One might have supposed France to have been sufficiently proud of Henri Pitot to have seen that his name was included in the roster of distinguished French scientists—like Coriolis, mentioned in the previous chapter—inscribed in gold on the main parapet of the Eiffel Tower. But since he died eighteen years before the Revolution, and since Gustave Eiffel favored those who had brought honor and glory to the newly made *République*, Pitot didn't make the cut, and his heirs have had instead to make do with just a side street named for him in the far southwestern town of Carcassonne.

{ 5 }

The pressing human need to know in as detailed and specific a fashion as possible the speed of any wind found its eventual champion in 1846, when the hero of this part of the story, the previously noted Thomas Romney Robinson of Armagh, created the first classical-looking modern anemometer. True, most of those earlier devices, employing such entities as paced-out fields and flags and vacuum tubes, were known as anemometers too, since they all performed the same function, for better or worse. But the Robinson anemometer, which is used to illustrate most encyclopedia and dictionary entries for the word, is the standard model—not least because it is so elegant and, in the way of things Victorian, so very beautiful.

It is made entirely of metal and is maybe fifteen inches tall—

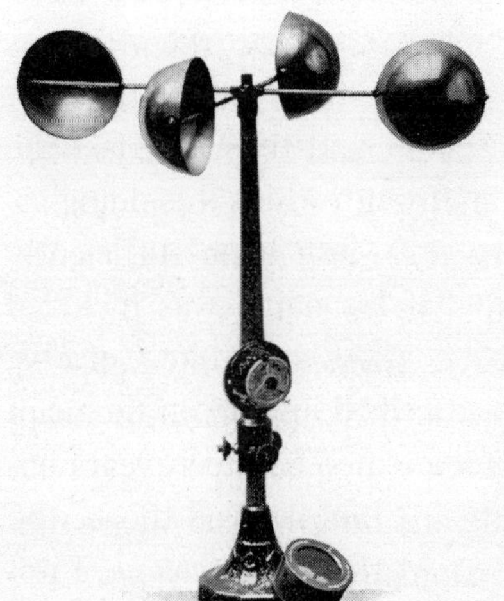

Thomas Romney Robinson's eponymous four-cup anemometer, with its rods and gear wheels allowing the wind speed to be read on the dial at the base, survives little changed almost two centuries after its invention in 1846.

designed to be placed up on a pedestal, out of reach of the gusty caprices encountered close to the ground. Rising from the sturdy metal base is a single hollow stalk, at the top of which are four rotatable arms, forming a cross, with at their far ends half-hemispherical metal cups, their meridians cut vertically such that any horizontally blowing wind could strike their open concave faces or, indeed, the closed convex backs directly opposite to them and set the upper part of the contrivance spinning around its vertical axis. The spindle inside the hollow tube would be connected through a simple gearing system in the base to a dial, from which the speed of the wind impinging on the cups up above would be recorded in miles per hour or meters per second or whatever units might be needed by the user. What could be more elegant, more pretty, and, to repeat the rhetorical question, more Victorian?

As this description hints, the apparatus known as a four-cup rotational anemometer, and as designed by the good Reverend Doctor Robinson, is—as is so much else that has to do with what outwardly seems such a simple entity, the wind—more subtly complex than it looks. For a start, consider the wind blowing in a straight line and hitting, as it will, two cups at the same time—one being open-faced and concave, the other closed-faced and convex. Surely, any reasonable mind might say, the impingement of the wind on cups on both sides of the spindle will cancel out and the cups will not rotate, the spindle will remain still, the gears below will not turn, the dial will read zero, and, since there self-evidently is wind blowing at the top of the device, the Robinson anemometer will have failed at its task.

But stay. Robinson's genius derives from the simple observable fact that to the approaching wind, the cups before it are of a profoundly different aspect and shape—one being scooped out and hollow, the other being smoothly and bulgingly semicircular. At this point enter the science of fluid dynamics—glancingly discussed during the earlier explanations of M. Coriolis and

Messrs. Hadley and Ferrel, *qv*—and the concept of the *drag coefficient.* This is a concept that states how much an object of any certain shape, size, and configuration exerts a drag on any fluid passing through or across or beside it at any angle. People who wish to go particularly fast—competitive cyclists, race car drivers, railway train designers—like to minimize the drag coefficient of whatever equipment they are using, keeping it to as low a number as possible. The drag coefficient of a jet fighter's wing may be 0.09; that of a B-2 stealth bomber 0.017, which is among the lowest coefficients of any man-made object. A dolphin can have a number as low as 0.0036, indicating the superiority of nature.

An open-faced hemispherical cup, as in Dr. Robinson's famous instrument, has a universally recognized drag coefficient (which Dr. Robinson would have known, since tables of the drag coefficients of all kinds of shapes were published widely by the 1840s) of a whopping 1.42. The cup on the other side, which presented its curved backside to the wind, exerted a much lower drag, of 0.38. Not exactly a stealth bomber, maybe, but allowing the breeze to slip fairly smoothly across its flanks.

The important number, though, is the difference between these two coefficients, high on one side of the spindle, low on the other—which means two things: that the spindle will rotate and the one cup will not cancel out the other; and the torque represented by this difference can be easily and arithmetically converted into wind speed. So, while the dangerously whirring cups will not be moving at quite the same speed as the wind, they will be moving in a direct *proportion* to its speed, and that proportion can be converted, by a metal-geared-down Victorian equivalent of what we today would better recognize as an algorithm, into the wind's actual velocity. A vane attached to the spindle and allowed to swing with the wind like the iron cow on top of my Massachusetts barn would tell the direction of the wind. Direction and speed, available in a single handy instrument; one-stop shopping personified, courtesy of the good Dr. Robinson of Armagh Observatory. QED.

The types and varieties of other wind speed measurers are legion. My own anemometer, set up high on the rooftop of my chicken coop, is a three-cup rotational model, which naturally (since the three cups are angled at 120° to each other, nearly obviating matters of torque and drag) has neither the technical advantages nor the cost of a four-cup version. It doesn't pick up the wind as comprehensively and is not so accurate. But it radio-transmits this speed and direction every two seconds to a receiver in my kitchen, so on a chill winter's night I can know how heavy my coat should be before I set out to the compost bin. I paid maybe $350 for my instrument; if you choose hot-wire anemometers or laser anemometers or those that determine velocity by way of ultrasound or acoustic resonance or wish to indulge in the caprices of particle image velocimetry, whatever that may be, you can spend small fortunes.

{6}

Or you may more wisely choose to set all complex numbers and decimal points and matters of calculus to one side and present instead a means of showing just what a wind actually does, what its effects may be, rather than reducing its majesty and moment to the cold and clinical universe of figures and formulae.

That is what Rear Admiral Sir Francis Beaufort did well over two centuries ago, in 1805; and while the peculiarly British star-struck world that admires initials awarded him the long tail of letters KCB FRS FRGS FRAS and MRIA to wear in the wake of his surname, it is for his much more simple scale of thirteen points—a single line short of a sonnet, it might be said—that has ever since brought an enduring kind of poetry to the wonderful, wild world of wind. The good reverend doctor of Armagh has faded long ago into the deserted uplands of history; the good admiral of Navan, County Meath, will likely be remembered always.

Francis Beaufort cut his teeth on the terrifying waters off Cape Horn. Those who sailed through these furious seas sometimes left memorials of their having done so. Beaufort left his scale; a century and a half later, an extraordinary New England sailor named Irving Johnson, who rounded the Horn as a young crewman on a four-masted German square-rigged cargo ship, the *Peking*, left a filmed record of the achievement. Mention of it is worth including here, since it gives an indication of the kind of experience that shaped young Beaufort's thinking about the two great forces, wind and water, that utterly dominate the lives of any who round the southern tip of the Americas, where the raging sea squeezes itself through the Drake Passage between the rocks and the icebergs of Antarctica, and endures—and survives—the worst of those winds named the Roaring Forties.

The *Peking* was the pride of Hamburg's Flying-P Line, in the 1920s quite dominating what was known as the nitrate trade, shuttling manufactured goods to South America's Pacific ports

Admiral Sir Francis Beaufort's name is forever linked with the scale he created, relating wind force to the observed conditions it creates at sea at various intensities. Other measurements, more mathematical in nature, have been introduced; none has replaced his classically empirical scale.

and hauling back thousands of tons of seabird guano to fertilize the farms and fields of Northern Europe. It was during this time that Irving McClure Johnson signed on for the simple pleasure, as he put it, of rounding the Horn "the wrong way," east to west against the wind, in a great sailing vessel. Captain Johnson went on to great things in later life, being in the US Navy and teaching youngsters how to sail, and to help in doing so owning three fully dressed vessels—a schooner, a brigantine, and a ketch—to show them the ropes.*

But as a contribution to history, perhaps Johnson's greatest achievement is having remembered, before the crew eased springs and cast off from Hamburg bound for Valparaiso, to pack his movie camera and a great stash of black-and-white film. This he eventually turned into one of the great classic documents of the maritime world. There is no footage like it, nor is there any narration remotely similar. He narrated it himself in the 1980s, entirely without a written script, the result having an immediacy and an urgency that makes watching it one of the supreme pleasures of a landlubber's life.

Here, for example, is the raw and unedited transcript—the captain, his unmistakably patrician tone melding with the timbre of his unmistakably New England accent, barely takes a breath as he tells of what he sees, his own remembered movie footage on the Mystic museum screen before him:

> Now watch this, just like the water coming over Niagara Falls. Look at this: the open ocean; the forces involved are fantastic, there's no

* Johnson and his long-suffering wife, Exy, sailed thousands of sea miles with his students, and on one voyage discovered in the surf around Pitcairn Island the anchor of HMS *Bounty*, the vessel commandeered in 1789 by the mutineers led by Fletcher Christian, who then settled this hitherto uninhabited few square miles of mid-Pacific lava and scrub. The anchor, along with Christian's Bible, remain in the tiny island museum, seldom visited—though I was fortunate enough to go twice.

words that I can use in any language that'll tell you what it's like. If you have been there, that's the only way you'll know, because the forces are beyond anything you've ever experienced or thought was possible.

The lower topsails are built to stand any storm, and yet we had trouble there too, but the screaming is beyond belief . . . She is really pumping and straining; the noise of the vessel, groaning one piece against the other, you think it was coming apart. But it's built for Cape Horn, and it's not about to come apart, and oh! the chafing and the wear and tear, not only on the ship but on the individuals on the deck. It looks like the ocean is running around, the ship is staying still; every line is just quivering and screaming and the lookout is back here and I'm on top of the chart house, and now the camera's held steady with the horizon, which is the proper place way to take pictures, but to get pictures from down here when a rogue sea may wipe you off is just asking for trouble, and avoid trouble came in a big way—all of a sudden I was afloat with a smashing crash, couldn't even shut off the camera as I was knocked gallywest. There she goes, smashed up completely and over the side; there was I washed off the top of the chart house, but then that's the kind of thing I'm dying to get pictures of to show my other brothers and sisters what it's like off the Horn.

The skipper on deck was blowing his whistle trying to make the man up top stop, but we couldn't even hear him five feet away—but when they get that clew of the sail pulled up you'll see a four-inch strip of canvas down the lower right-hand corner just hanging down there, and you just barely see it now, but the third mate gets up on the end of the yard, could not reach that strip of canvas so he hung down and I missed the best picture of the whole trip—I ran out of film; he hung down by one arm and one leg from the foot row and just then a squall came and the ship rolled and that man's entire body vibrated just like those sails, just a whole body just flapped just like the sail; never have we ever heard of or seen such a thing as a man's body flapping like a sail, but when the man came down nobody said one word

> to him. We all would have done it; some kind of hypnotism makes us do those things. This is just incredible when you look at them later; it's, it's, it's crazy but they do it on the conditions of working for the ship, making it go. She needs you, and when there's something like that going on, you would do the same thing too when you're young and under those conditions.

{7}

That was in 1929. Some 120 years before, Francis Beaufort experienced such things on many a great square-rigged sailing warship too, and regularly. Irving Johnson turned what he saw and experienced on the *Peking* into a memorable morsel of cinematic history. Francis Beaufort turned his remembrances into an enduring trope of meteorological poetry—and he did so after a long and varied experience, not on giant four-masted German cargo vessels bent to the nitrate trade but aboard a succession of sturdy three-masted Royal Navy warships, most usually the kind of heavily armed frigate known as a *man-o'-war.*

It is said that he craved a sailor's life almost from infancy, and duly signed on his first ship at the age of just fifteen, supposing that an apprenticeship aboard the privately held East Indiaman, the *Vansittart*—which had orders to combine Asian treasure-seeking and trading with a measure of coastal surveying—might help advance his chosen naval career. It did so indeed, though a baptism of fire awaited him: he was shipwrecked in the Dutch East Indies (quite congenially, as it happened, his captain mistakenly driving the vessel onto a coral spike and allowing it to settle gently onto a reef). Young Beaufort acquitted himself memorably, and in a manner that impressed all who heard of him when finally he arrived back in London after two years away: he had superintended all the timekeeping operations aboard the *Vansittart* during her floating months, had corrected the coordinates on the poorly made Dutch charts of the waters off Java, and when his rescue ship brought him unexpectedly to a port in eastern China, set to surveying and writing with a vengeance, returning home with intelligence of considerable value to traders and sailors alike.

Then came the Navy—England's imperious senior service, based, as Churchill would later say, on the unholy trinity of *rum, sodomy, and the lash.* And nepotism, he might have added, which played an undeniable part in Beaufort's subsequent career—his Anglo-Irish family was well-connected with many of the senior lords of the Admiralty—and after that first ill-fated commercial voyage, Beaufort's career with the Royal Navy, fraught with difficulties though it may have been, was now firmly set. Faraway service on frigate after frigate followed—HMS *Latona*, HMS *Aquilon*, and HMS *Phaeton* most memorably, each vessel seeing plenty of action against the Spaniards, the Dutch, and, of course, the French, in which Beaufort won his naval spurs, as it were, and took wounds aplenty for his pains—nineteen of them during one Mediterranean encounter alone, some of them from shot and shell, others from the broadswords used in on-deck skirmishing (with swinging ropes, rigging-clambers, boarding attempts) that were characteristic of naval battles of the day.

He won a wounded sailor pension of £45 a year and was let go from service for a while, and then posted, eventually as commander (his first such assignment) to a mostly stay-at-home vessel, HMS *Woolwich*, officially described as an armed store ship. He made one quick turnaround trip to Bombay to evacuate soldiers wounded in a local conflict, then returned to London—Deptford first, and the Thames-side dockyards at Woolwich from which his vessel took her name—before making his way down around the Kent headlands and into the Channel to Portsmouth, there to wait orders. To wait, and wait, and wait—week upon week, month upon month, with Captain Beaufort, by now, in his increasingly acid view, overlooked and underpaid. Though a friendly and well-balanced man by most accounts, he vented his steadily increasing fury—privately, in his leather-bound journal—at the senior men of his service. His was of a tradition all too familiar to soldiers and later to airmen alike, to be memorialized a hundred years later in the Crimea by the Russians commenting on the defeated British army as being composed of *lions led by donkeys.* To Beaufort, spitting rivets in his master's cabin, the lords of the Admiralty were the donkeys, his leonine sailors and himself insatiably eager to get going, to do what sailors were designed to do—to *sail.*

It was during this period of melancholic frustration, during the dark English winter of 1806, trapped, as he saw it, in the naval anchorage at Portsmouth, that the usually sweet-tempered captain calmed himself by writing what would become one of the more consequential lines in meteorological history.

Already he had voiced his private disdain at the manner by which wind and weather were recorded and forecast by the naval establishment of the day; he had a particular loathing for numbers, for the widespread affection for reducing weather to mere arithmetic, which meant little or nothing to him or to his sailors. Hence his venting—this time not so much at his Admiralty superiors but rather at what he saw as a deeply flawed system. It was January 13, 1806, and he was sequestered in his captain's cabin at the stern of the *Woolwich.* What he wrote that day appears, legible to a fault, in his private log (as distinct from his official commander's log, each volume of which by law had to be sent to the Admiralty at six-month intervals, for the archives, with a fresh one started in its stead):

"Hereafter I shall estimate the force of the wind according to the following scale, as nothing can convey a more uncertain idea of wind and weather than the old expressions of moderate and cloudy, etc., etc." He then wrote the list that follows:

0. Calm

1. Faint air just not calm

2. Light airs

3. *Light breeze*
4. *Gentle breeze*
5. *Moderate breeze*
6. *Fresh breeze*
7. *Gentle steady gale*
8. *Moderate gale*
9. *Brisk gale*
10. *Fresh gale*
11. *Hard gale*
12. *Hard gale with heavy gusts*
13. *Storm*

The Beaufort scale was beginning to take shape. The elegance and beauty of what the young sailor would then develop over the coming months and years—and it took a full quarter century before it was refined, accepted, and then wholly adopted by maritime watchkeepers everywhere—is that it was founded not on numbers but on *effects.* What did the wind do to the surface of the sea at each incremental increase in its power? And, more significantly, what did it do to the heavy canvas of the sails, the dozens of square yards of stiff and, one hopes, unbreachable fabric that would fill with wind and drive the ship onward through the sea?

Sails, their appearance, form, and function, demonstrated the practical side of the scale; Beaufort's spare and, over the years, further refined description of the sea's appearance offered up, at least in calmer moments, the poetic.

Consider the *Woolwich*, trying to nose her way out of the Portsmouth roads and into the broad waters of the Channel, bound maybe for Montevideo or Martinique or Guadeloupe, all places she would indeed soon visit, with bellicose imperial purpose, once her captain received his eventual sailing orders. But at first there was no wind at all.

When windlessness was all there was, the surface of the sea

The sea states photographed at a pair of Beaufort scale numbers—the Flat Calm of Beaufort Zero up to the Strong Gales of Force Ten. At the two still higher forces than this, photography is nearly impossible, with the sea state at Force Twelve officially described as "phenomenal."

was quite flat, like mirror glass, its waters unruffled by any movement of the air above it that might disturb the tranquility and reflective nature of its uppermost layer. The sails on the *Woolwich*—or indeed on a three-masted frigate of the first rate, the kind of ship that young Beaufort would have dreamed of commanding in those heady eighteenth-century days when Britain had unchallengeable control of the seas with her fleet of vast size, speed, and magnificence—were in the main square-shaped, suspended from the crossbeams or yards of the three main masts. When there was no wind whatsoever and the air was utterly still, they hung limp and impotent from their yards. The ship would not, could not, move; and since the three cardinal attributes of any warship, inscribed on every bridge even to this day as *vade mecum* for her captain, are to be able to FLOAT, MOVE, and FIGHT, when the wind is still there is nothing to do but FLOAT, and set your sailors to the mundane tasks of holystoning and painting and peeling potatoes.

But then, seemingly out of nowhere, and felt on every cheek of every scurvy wretch idling out on deck, there comes a sudden cat's-paw, the gentlest puff of wind. The reaction is immediate and universal: To a man, the entire ship's company on instinct looks upward, craning their necks to consider the shape and situation of each and every one of the sails. On a frigate of the day there would have been twenty-eight such canvas sheets—eighteen of the square-rigged kind, with six on the foremast, seven suspended from the mainmast, and a further five on the trailing mast, the mizzen. Then, and in addition, fore and aft, would have been the triangular or otherwise eccentrically non-square sails, the jib sails rigged between the foresail and the bowsprit, and then, near-invisibly snugged between the sails on the mainmast and the mizzen, would be the spanker and the aft staysails.

And all these enormous sheets—a mainsail might weigh a full ton, and when wet or coated with ice or packed with wet driven snow, incalculably more—were quite without function in a Beaufort scale zero calm. But with this sudden new lick of a breeze, maybe one or another of the sails would show a sign of life; hence the sailors' upward gaze, to see which, if any, might respond.

Invariably, it would be the uppermost smaller sails, those seventeen stories high on a well-found vessel's mainmast, that would start to flutter. A cry might go up from below—*The royal's moving! The skysail's showing life! Look at the t'gan'sl going!*—the topgallant sail rendered into the language of the common sailor. But the ship herself would behave with sturdy indifference, not deigning to shift herself through waters that were, however, no longer mirror-flat but had turned fishlike, scaly, the surface pocked with the slightest of ripples. A light smack casting her nets on such waters might well be able to move and turn in so light a breeze, but the possibility of these hundreds of tons of heavy wood and iron of the *Woolwich* or any other such square-rigger enjoying the privilege of steering way remained nil. This was Beaufort scale One, *Faint air just not calm*, as he called it. To a big ship, a real ship, almost useless.

Ratchet up one further notch, to Scale Two, *light airs*, and things aboard the frigate begin to change quite noticeably. A series of audible thuds can be discerned from up above. One by one, from the royals down to the mainsails and the courses, the sails begin to bow open, filling now with a real and actual wind. The vessel begins to move over water that is now stippled with small wavelets, none of them breaking and the surface still with a glassy appearance. The steersman turns the wheel and the ship responds, starting to turn as the wheel and the rudder directs, the wind at last giving him steerage way, the compass rotating in the binnacle, the deck at the moment of turning displaying just the slightest lean. And there is now a definable windward side to the ship, the sailors feeling it, the waters on the hull's lee side calming, protected by the shroud of the passing vessel.

And back on shore, some leaves start to rustle, weathervanes begin to swing, people standing stationary outdoors can feel a cool wafting against cheek and jowl.

Now up to Beaufort Three, *a gentle breeze,* and then a close-hauled ship with her sails tight and held fast by her numberless halyards, stays, braces, and lifts will start to move with dispatch and purpose. Once she gets going she can now make maybe four knots, her bow bubbling slightly as it cuts through a sea that has a visible sparkling liveliness about it, with a few of the wavelets' crests now breaking, leaving tiny flecks of foam. Forces Three to Four provide the beginning of a good working breeze, ideal sailing weather for ships big and small—even the fishing smacks visible on the horizon are now careening noticeably and are skimming along, making fine headway.

By Force Five, *a moderate breeze* has become a *fresh breeze,* and a warship's captain has to begin to make a new set of decisions, both to take advantage of the wind's gathering power and to prepare precautions, just in case. The sails are quite taut at this point, and the rigging is beginning to thrum with an audible *basso continuo.* The sea is generously occupied by waves, longer and more numerous, many of them breaking white, with some clouds of spray visible from up on deck. And by the time Force Six is reached—the point adroitly designed by the admiral to be the halfway point of his scale—the officers are gathered on the bridge, deciding whether now is the time to shorten the sails, since the vessel is positively flying along and is leaning hard, and, like a horse extending its stride and pace from a canter to a full gallop, is becoming just a little harder to handle.

Between Forces Five, Six, and Seven—as the winds strengthen still more, leaving behind the rather benign term breeze (of Spanish origin) and acknowledging the appearance of the new generation of more serious winds, of a kind employing the term *gale* (maybe from the Old Norse for *mad*) in a variety of iterations. Now decisions have to be made, certainly in a sailing vessel, since wrong choices at this point could have perilous consequences.

As the captain recognizes the forming of Force Seven by examining the state of the waters around him—*the sea heaps up, white foam from breaking waves begins to be blown in streaks*—he orders his crew to climb above into the yards and begin the process of steadily and methodically shortening sail. They do so by tying reefs—light lines attached to the sails—in four increasingly restrictive stages, a single reef reducing the sail by a quarter, a double halving its exposure to the strengthening wind, then a treble reef, and finally, when the wind begins to shriek and the ship clearly has so much wind bearing down on its remaining exposed sail area that it will soon become dangerously unmanageable, the fourth reef, creating a situation known as a *close reef*, where essentially the entire sail is snugged away to the yard so tightly—and this is not just one sail but all of them, on all three masts—that it ceases to be, in terms of its effectiveness, a sail of any sort at all. A very few storm jibs are kept flying to continue to give the storm-beset vessel some steerage way, so that she can at least try to maneuver her way out of the gathering maelstrom.

Meanwhile, the scale numbers continue to climb, and other, lesser sailing ships in neighboring waters scurry for the safety of a nearby harbor or else clear all coastlines and head out into the deep sea, lying-to under bare poles. In so doing they are purposefully allowing the ocean and the wind to toss them around like corks in a waterfall, uncomfortable for all aboard but, unless embayed or driven onto a lee shore, safe from a skipper's worst enemy in a storm—the viciously sharp rocks and grounding perils of an approaching coastline, the dangers of the sea fast overtaken by the perils of the land.

It will be during a Force Ten event—designated by Beaufort simply as the outward manifestation of a *Storm*—that a prudent captain will have ordered his sails reduced to almost nothing. He will have no firm idea of how fast the wind might be blowing; no anemometer will have told him anything with a measurable numerical value. But he will know instinctively what he is up against.

If he is as wise and experienced an old salt as a big square-rigger demands, the state of the sea around his embattled ship will give him the general idea: *Very high waves with overhanging crests* will be frighteningly evident. *Sea takes on a white appearance as foam is blown in very dense streaks; rolling is heavy and visibility greatly reduced.*

So he will order his bosun and first mate to demand the sailors come down from the rigging, certain that all the sails are safely clewed up and his men safe and sound below, since there is little more they can do.

Beyond this, the *Violent Storm* and the *Hurricane*, Beaufort Forces Eleven and Twelve, respectively, were seldom experienced back in sailing days, and are avoided at all costs by prudent mariners of all stripes today—though it has to be said that the possession nowadays of stout iron hulls and powerful engines and radio transmitters and radars are likely to keep today's traders and warriors well able to FLOAT, MOVE, and FIGHT with much greater ease than back in Beaufort's day.

{8}

Beaufort's is a scale of thirteen points, zero to twelve, which makes impeccably good sense, universally. Though it has an exact midpoint—Force Six—it also has an inflection point—Force Seven—beyond which ships and humans both start to struggle with a world of wind that up until then has been, in terms of the experience of it, generally benign, maybe even fun. But at the point where it is more serious than making it tricky to open an umbrella or to light a match or to fold a broadsheet newspaper while sitting on a park bench at crossword time—then wind starts to become a more malevolent entity, merging its accelerating magnificence with a growing degree of danger. It becomes difficult to stand upright. Hats blow off and are set to bouncing down the street. The air stings the face, especially if it is cold, laced perhaps with hail or snowflakes or frost. Cyclists are obliged to set down and walk. Doors slam open or shut. Windows rattle in their frames. Ships' bows plunge or rear into and out of the swells, lifelines are clutched, and in ocean liners unbolted grand pianos skitter across ballroom floors made slippery with French chalk. All such things would begin to happen back at Force Seven and above. By Force Twelve, however, the piano would be no more than matchwood, and no glass would remain in any but a ship's porthole, and vessels out at sea in such conditions—the official word for the sea state at Beaufort Force Twelve is, appropriately, *Phenomenal*—are careless of all mundane matters except for how best to keep the engine going and the rudder working so that the craft can keep head on to the sea and the gale, for being abeam of either would mean a certain capsize.

And it could be even worse, especially if a whirlpool is involved. Between the Hebridean islands of Scarba and Jura is the Strait of Corryvreckan, notorious for its tidal races, its howling westerly gales—and a whirlpool that has been the stuff of legend for centuries. Michael Powell and Emeric Pressburger set their classic 1945 romance film *I Know Where I'm Going* there, with our heroine Joan Webster beset by inclement weather and so quite unable to meet her intended on the remote island of Killoran. There is a memorable wind-related scene, when Ruairidh Mhór, the weatherbeaten old Hebridean boatman, tells his impatient visitor why she may *not* get across the strait to the Isle of Killoran in the teeth of that day's gale. *How long will the gale last? As long as the wind blows, my lady. But it looks so close—in half an hour we could be there. In less than one second you could get from this world into the next . . .*

IT WAS TO take some long and seemingly interminable while for Francis Beaufort to have his scale—which he refined and refined as he wandered the world in a bewildering variety of vessels—wholly accepted, despite its very obvious brilliance, its grace, and its intelligence. He was to go on to greater things indeed. He became adept at surveying—an occupation that he had taken to back in his teenage days on the *Vansittart.* He spent some while, for instance, surveying the southern coast of Turkey, Smyrna to Bodrun, Tarsus to Iskendrun, a coastline peppered with fantastic antiquities, on which he would become something of an authority. During his travels in Asia Minor he was set upon by brigands and badly wounded, and was obliged to return home, where he wrote a book that sold well and made him something of a household name. Though not, quite yet, for his scale. That would have to wait until 1829, when he was finally promoted to the office he had secretly coveted for years, that of Hydrographer of the Navy. Already he had hand-drawn charts of the Turkish coast that were so meticulously made that they would go from his hand straight to the engraver. As hydrographer, with dozens of survey ships ranged across every ocean under his command, and with a worldwide appetite by traders and warriors alike for accurate and up-to-date charts, he achieved fame on bridges and in boardrooms across the entire world.

And ironically, it is not for his scale but for this beribboned elevation—his twenty-six years as Britain's national chart-maker and oceanic depth-finder—that Francis Beaufort is still officially best-known. He deserves accolades, no doubt. During his tenure he published nearly seventy charts a year—by the time he retired from the Navy in 1855 he had more than a thousand to his name. To some still today this is by far his most significant achievement. The *Oxford Dictionary of National Biography*, placing its formal imprimatur on Francis Beaufort's life, writes almost disdainfully of the reason for his popular appeal: *Although the public connects his name only with his wind scale, among seamen and especially hydrographers Beaufort's achievement is remembered with awe.* In the entire entry, three full columns long, that is the sole mention of the scale.

And yet, late each British night, tens, maybe hundreds of thousands are still lulled to sleep by the near-mystical poetry of the *Shipping Forecast* broadcast on the BBC. And here the scale is used in a determinedly casual way, the assumption that everyone listening knows perfectly well the highly nuanced code in which it was written more than two centuries before.

Fitzroy, the announcer may say, referring to the name of a sea area far off western France. *Cyclonic six or seven, later become variable three or four, then northeasterly five or six.*

And those abed will know—will know what it means for a sailor in a rotating storm in the eastern Atlantic Ocean having to face high winds and big seas that mercifully will abate during the night, only to pick up again, cold and strong and coming from the northeast, the very next morning with swells and breaking waves and the low whistle of a near-gale audible up in the rigging. Snug and warm in bed, though maybe with a winter rain tapping at his windows, the listener will turn over and close his eyes and the words will fade away, Six and Seven, Three to Four later, then calm. And by then he will be fast asleep. While out at sea the skipper on his rolling, heaving little craft will be alert and on the bridge, listening out, wanting to know what Admiral Beaufort's unseen forces have in store for him on the coming dawn, and hoping the scale number will be comfortingly low and on its way downward, and that the seas will thus moderate and steady themselves and the rolling will become more gentle, the barometer will rise, the visibility will improve, and his homeward journey to his wife and children will be more certain, becoming safe and sound.

Meanwhile, other winds for other people are all still everywhere, far away from Fitzroy, waiting in the wings. Gentle winds that disperse; stronger winds that do work; mighty winds that destroy. Dispersing, doing, and destroying—the ceaselessly cycling trinity of moving air, essential and eternal, and assembled into categories by a long-passed Irish admiral. They are presented as culminating in the pages that follow here, the raging and inclement storm winds, offered up in just the order that the good admiral long ago suggested.

CHAPTER FIVE

Inclement Winds

Listen. Listen! **That's some of what done it—the dusters. Started it, anyway. Blowin' like this, year after year—blowin' the land away, blowin' the crops away, blowin' us away now.**

—Muley Graves, Oklahoma farmer in the 1940 movie version of John Steinbeck's *The Grapes of Wrath*, dir. John Ford.

{ 1 }

It was the front-page lead story in *The New York Times* on Saturday, May 12, 1934: "Huge Dust Cloud, Blown 1,500 miles, Dims City 5 Hours" read the headline in column one. The dust, soft and gray when it settled, dirty yellow when lit by the fitful sun, black when seen from above, had arrived without warning around breakfast time on Friday—though airline pilots had told their control towers some short while before dawn of seeing it approaching, an enormous and seemingly impenetrable slab of dark grey in the sky, moving unstoppably eastward, and that they had to climb up three miles to avoid it.

From then until after the Manhattan office lunch hour, New Yorkers could barely see across the street; a thick haze enveloped everything. Streetlights went on and cars had to switch on their headlights, as if in an eclipse. Tourists had hoped for a fine view from the top of the newly opened Empire State Building, the early summer's day having dawned so crisp and clear, but they were sorely disappointed: nothing could be seen of the city below, whether they looked south in the direction of the Statue of Liberty, north to Central Park, or east toward the Brooklyn Bridge; all was blanketed in a miasma of darkened cotton wool. Out at sea it was no better: a freighter inbound from Germany had to turn around outside the narrows, as the city, so easily spotted from out at sea, seemed suddenly to have vanished—and back in 1934 there was no radar to provide the captain with reassurance that it was still there.

The cloud then all but vanished by mid-afternoon, heading south to dump yet more dust on Washington, DC, before pushing out into the Atlantic. Radio stations in Chicago, Detroit, and Cleveland all reported that the very same cloud had visited them during the night hours and late the previous day, confirming what all in New York City suspected. There was little doubt as to the origin of the dark-yellow dust. The skipper of the *Deutschland* had an idea: when he finally got to berth he remarked that he had only seen anything like it at the Cape Verde Islands in the eastern Atlantic Ocean, all too frequently covered with windblown dust from the Sahara desert. This was much the same, he thought; it was soil from somewhere else, deposited without asking on the people of New York City.

He was exactly right. According to the *Times*, the falling dust "used to be topsoil from the valleys of the Missouri and Mississippi Rivers." Two hundred million tons of it at least, a mass of once cultivable and usable prairie topsoil now all blown up and eastward, clear across the country, by powerful northwesterly winds. This was indeed visible, palpable, irritating, and troubling evidence of the phenomenon that would come to be known as the Dust Bowl.

The phrase, which suggests a geographical and topographical entity, was in fact used to denote a series of terrible events, of catastrophic dust storms and droughts that over five wretched years had become the rural expression of what was underway across the entire nation, and that was the Great Depression. Not only was it not a geographic expression but it had little to do with bowls—quite the reverse, in fact, involving what was topographically speaking more like an immense dome of land making up the High Plains of Oklahoma, Texas, Kansas, Colorado, and Nebraska, once called the Great American Desert. The events were known but slightly by most outside the High Plains—but now here in New York City on that springtime Friday it had for the first time come into sharp focus, had been brought dramatically eastward, almost as if the embattled and desperate people of the prairies 1,500 miles away out west were mounting a last-ditch effort to gain the attention of those in power in eastern America, to shake them awake, to maybe compel them to give much-needed help with their windborne plight.

The slowly unfolding disaster that overwhelmed these prairie settlers had started early that same year, and it struck again in 1935, 1936 (when the term *Dust Bowl* was first officially employed), and 1939. The images are horrifying and legion: giant rolling clouds of black dust, half a mile high, thundering out of the west and bearing down on farms and fields and fences that look fragile and inconsequential in the face of this intense barrage of nature at its worst. Rolling ahead of the duster itself, the air is often charged with static electricity, powerful enough to knock out a car's engine, to smash a man off his feet, or to electrocute a child. A lost child might also suffocate in the dust; cattle and horses, half-starved because the crop-sustaining soil had been blown away, with teeth broken by their desperate chewing at what gravel remained underfoot, would go hysterically mad with the dust being rammed into their nostrils. It was all about the wind-borne dust that was choking and smothering all the homesteads and sharecroppers and villages that lay in its path—dunes nine feet high would engulf a home up to its shingles, burying the outhouse completely, leaving only the root cellars where one might find the skinned carcasses of bludgeoned rabbits and canned tumbleweed, the only foodstuffs that the more prudent and tenacious settlers had tried to store for when hard times became impossible times, when they were marooned by the horrors of the dust and by the "black blizzards" of winter, when black snow and grit created their own kind of dunes every bit as impenetrable as during the rest of the year, but which then froze solid, like ice-cold concrete, entrapping whole towns.

These wind-borne assaults are the infamous *dusters* of John Steinbeck's writings and this chapter's epigraph, taken from Nunnally Johnson's screenplay for the John Ford movie of *The Grapes of Wrath.* For what Steinbeck described was how the dust storms of the thirties had quite literally blown people off their lands, had driven them out of their shabby little sharecroppers' farms, out of their counties, out of their states, forcing them—ten thousand of them every month at its worst—to drive their battered jalopies westward to the far ocean states, for the fields of California and Oregon and Washington, where they might find hope and salvation and better times but instead found yet more poverty and

A classic "duster," a windblown cloud of soil that buries entire farms, chokes livestock, and renders entire prairies uninhabitable and unworkable. The bane of the Dust Bowl years that further depressed America's idea of herself during the Great Depression.

misery, in a country gripped by a Depression of which the Dust Bowl was just the most powerfully symbolic manifestation.

It would be idle, however, to blame the wind entirely for the tragedy. Man—the white man, exclusively—was by far the more responsible. The Native Americans who had lived there alongside the bison herds long before the murderous settler colonists arrived—the Cherokee and Choctaw of Oklahoma; the Comanche and Apache of north Texas; the Osage, Pawnee, and Kiowa of Kansas; the Navajo of New Mexico; the Cheyenne and Arapaho of Colorado—all had suffered regularly from the vagaries of weather and season, from the droughts and windstorms that habitually plagued the High Plains. But their trials were wholly manageable, and the nomadism and economy of settlement that were central to their styles of living allowed them to survive and bounce back and even to prosper in a modest manner. But everything changed for them and for the land when the government in Washington decided, with the Homestead Acts, to populate the plains, spear the railroads across the landscape, turn a blind eye to the willful slaughter of the bison, and introduce technologies to the wretched business of small-scale dry-land farming, and that happened to be wholly unsuited to the environmental requirement of the territory.

Steel-bladed plows—those made by John Deere have passed into legend as the villains of the piece, though there were many other equally culpable tractor makers—ripped away the topsoil, tore out the grasses that had bonded the earth together, and then saw crops planted that might well bring the sharecroppers some sense of instant prosperity but in whose sowing lay the seeds of ruin and disaster. The government was much to blame. Speculators were much to blame. Thoughtless sloganeering like the infamous nonsense that "the rain follows the plow" was much to blame.*

---

* A deluded—and nowadays derided—branch of nineteenth-century climatology held that the simple act of plowing the desert and planting crops would have a

Agricultural carelessness, settler impatience, and speculative greed were much to blame. True, the wind was the entity that blew the ruined soil away and eventually helped drive the unfortunate victims out, and so to that extent the wind was indeed the proximate cause of it all. But the ultimate causes were quite otherwise.

But given all that, what was the possible cure? President Roosevelt, newly installed in the Oval Office but regarding the plight of the Midwestern farmers as paramount, turned for advice to the son of a Carolina cotton farmer, Hugh Bennett, who was an ardent and outspoken believer in soil conservation. Bennett had been railing for years at the irresponsibility of High Plains settlers in merely scraping away the soil and destroying the grass that had bound it together, and from an obscure position within the government's Department of the Interior tried gamely to make the nation listen. As it happened, the dust helped. The enormous cloud that enveloped New York City that May Friday in 1934 swept down to Washington, DC, over the weekend, the white marble city covered in gray grime at the very moment Hugh Bennett was testifying before Congress about the need for a Soil Conservation Service to be given power, and teeth, to enforce new farming methods. Bennett, ever the showman, had only to point out the window—"There goes Oklahoma, gentlemen"—to show the lawmakers the gravity of the crisis. Thus, and effortlessly, he got his way; and from that moment on, Roosevelt saw Bennett as the potential savior of the High Plains situation, the man who could bring the Dust Bowl to its end. If only the wind could be made to cooperate.

---

permanent beneficial effect on the local weather by inducing rain to fall on hitherto arid landscapes—*permanent* being the key word. The theory was employed by proponents of Manifest Destiny, who argued that Americans had a divinely inspired duty to expand westward across the continent, populating as they went.

{ 2 }

And the solution to that was down to Roosevelt himself, and a plan as outlandish as it was imaginatively ambitious. It was, he would always later say, his Big Idea—and this from a man who had wired most of rural America for electricity, dammed the Tennessee River, created Social Security, and constructed the atom bomb. To solve the problem of the Dust Bowl and stop the unstoppable wind its tracks, he would have his New Deal government *plant trees*—millions upon millions upon millions of them—in a series of long protective lines of windbreaks stretching from the Canadian border down to central Texas. Government agencies

The Paul Miyamoto oil painting Home *(2024) depicting an imaginary assault on domestic peace and tranquility as a formidable windstorm bears down on the property.*

that were dear to FDR's heart, agencies that he had set up in his frenzy of first-hundred-days activity, would participate: the Forestry Service would design it, choose the most suitable trees to plant (government scientists were sent to the Gobi Desert in China and to the Sahara in Africa to select suitable species), decide how best to plant them—in rows going north-south or east-west, how far apart the rows should be, how far apart the trees should be. The Works Progress Administration, the WPA, would then liaise with the various states to decide where shelterbelts were most needed locally, which farmlands were in greatest peril, which gale-blasted ridges should be clothed with new foliage and which would best be left undressed. Some of the three-hundred-thousand-strong army of young, single, and hitherto unemployed but now uniformed and fully employed men of the Civilian Conservation Corps, the CCC, would do the heavy lifting; they would

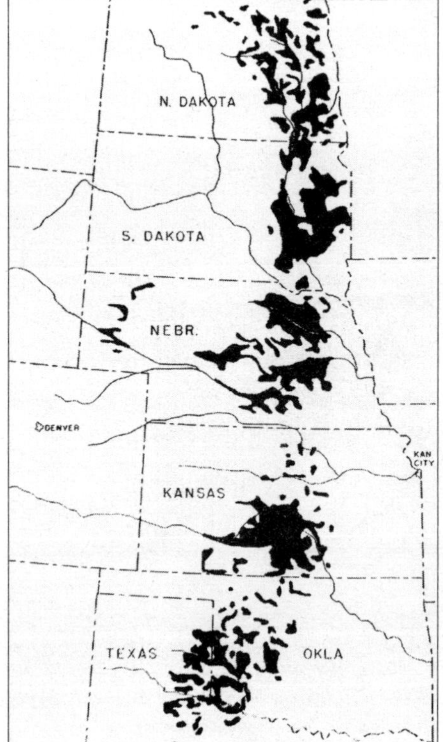

President Franklin Roosevelt ordered the planting of more than 220 million trees to create the Great Plains Shelterbelt, extending from North Dakota to West Texas, in an effort to curb the destructive power of the soil-eroding gales. The program began in 1934 and effectively continues to this day, with farms in the Midwest now generally prosperous and stable.

fan out from their new headquarters in Lincoln, Nebraska, and dig the holes and set the saplings in place, secure them to anchors by twine, then tamp down the topsoil and water each infant tree and keep an eye on things until all seemed secure and strong enough to let nature do its worst. An eleven-man CCC team (the men were paid a government dollar a day, with a quarter of that sent to their parents to spread the wealth around) could plant six thousand trees a day.

It had to start somewhere. It took a little finding, but beside a small country road outside the barely noticeable panhandle town of Willow, Oklahoma, there stands today a historical marker, almost entirely ignored by those few who pass it by. There must be tens of thousands of similar markers spread across the American landscape, noting the sites of battles or notable inventions or birthplaces of forgotten people. This particular marker commemorates a tree—and not, despite the community's name, a willow tree but an Austrian pine tree, *Pinus nigra*, which was planted here at the specific command of President Roosevelt. The marker explains:

> . . . his Prairie States Forestry Project envisioned planting wide belts of trees from North Dakota to Texas to protect cropland and reduce damage to the environment. The nation's first shelterbelt was planted on the H. E. Curtis farm near Mangum in Greer County, Oklahoma. Oklahoma's first State Forester, George R. Phillips, planted the first tree on March 18, 1935. From 1935 to 1942, 223 million trees were planted in 18,599 miles of shelterbelts throughout the Plains states, with 2,996 miles in Oklahoma. The nation's number one shelterbelt is located five miles east and one mile north. Like many of the original shelterbelts and the narrower windbreaks planted since then, it continues to provide conservation benefits to this day.

"This day" was 1995, when the marker went up. In the three decades since, many of the old belts have vanished, and the trees of which they were composed—cottonwood, honey locust, ponderosa pine, walnut, Chinese elm, ash, hackberry, Austrian black pine—have been gnarled and broken by age and time. But they did the job. FDR's Big Idea paid off. The High Plains of the central prairies, the land between the hundredth meridian and the front range of the Rockies, is no longer a dust bowl. It has grass once again, and in a few scattered places there are bison again, and wind pumps creaking and bringing up water to farms of moderate prosperity. But the Comanche and the Kiowa and the Arapaho are largely elsewhere now,* and farming of a very different scale and kind is underway—hog farms and cattle feedlots and oil wells and rare earth mines and military bases bring in most of the tax dollars and provide most of the employment. There is still wind, though—but instead of blowing dust, it blows through immense new wind farms, and provides electricity for firms like T-Mobile, Google, and Anheuser-Busch, brewer of Budweiser beer. A century ago the High Plains survived, barely, in spite of the wind. Today they prosper, largely, because of it.

{ 3 }

Over in Europe, the ill winds can be more powerful than these. The dusters of the prairies were not excessively strong, as winds go; their disastrous doings derived from what they tore up and carried with them, rather than any trees they might have blown down or houses demolished. But in Europe, many of the displeasing winds that are famed and named are dangerously strong, gale-force and flag-shredding and white-horse-generating torrents of invisible madness. And their names are all too well-known by those who live under their malevolent influence.

* Though at the time of writing the local congressman, Tom Cole, is a member of the Chickasaw tribe who studied British history at Yale.

Naming winds seems not to have been so much of a sport in the English-speaking world. There are some four hundred winds worldwide; in North America, only nine of these—the Duster and the Diablo, the Chinook and the Santa Ana, the Alberta Clipper and the Pineapple Express, the Nor'Easter and the Tehuantepecer and the Barber—have been accorded names; in Australia only five—the Brickfielder and the Freemantle Doctor, the Southerly Buster, the Black Nor'Easter and the Guburra, famous up in Darwin for signaling the end of the season that North Australians call *the wet*; in Britain only one sad little gusty thing gets a name: the Helm wind, a chilly easterly that blows over the Pennine Hills in Cumberland and occasionally leaves a helmet-shaped cloud over the summit where it collides with warmer airs on the west.

Leave the English-speaking world, though, and you soar into an abundance of winds with vernacular names. Some are familiar, or vaguely so: the Sirocco, the Buran, the Ghibli, the Haboob, the Harmattan, the Levanter, the Prodrome,* the Simoom, the Sonora, the Suhaili, the Tramontane, the Williwaw, the Zephyr. Most others are names seldom heard outside the regions where they blow: the Waryaraik of Argentina, the Uberre of Neuchatel, the Tebbad of Turkestan, the Stikine of Alaska, the Orkan of Norway, the Hawa Janubi of the Arabian peninsula.

And then there are two—ill winds both—that are extremely familiar to people living in and around the Mediterranean, and that each cause people to behave in unusual and unpredictable

* British embassies and consulates around the world used to have "Prodrome" as their telegraphic address—thus "Prodrome Buenos Aires" or "Prodrome Sofia" would get your message to diplomats in Argentina or Bulgaria with just a few words. The literal Greek meaning of the term denotes a light wind that precedes the rising of Sirius, the dog star. Britain believed her embassies were in the business of intelligence gathering, of gaining the first faint signs of some development within a foreign country. The light wind that suggests the coming of a storm . . .

ways: the Mistral of Provence and the Bora of the Italian city of Trieste. Taken together, and set in context between a moderate American wind that once wrought disaster and a truly fierce maritime wind that unexpectedly did the same, and lethally, this trans-Mediterranean pair ratchet the very notion of dangerous winds up a couple of notches.

{ 4 }

The Mistral, the cold wintertime scourge of the city of Marseille and the region of Provence, is a creature born, in common with most classical winds, of a unique combination of climate and topography. It requires the existence of two opposing pressure systems—one high pressure, operating in the Bay of Biscay, and one low pressure, working its way across the North Sea. The high-pressure cyclonic wind, settled to the north of Spain and the west of France, will be rotating in a clockwise direction—this always being the case in the Northern Hemisphere—while some three hundred miles to the northeast, the low-pressure area between the English coast and the offshore islands of the Netherlands and Germany will see its airs spiraling in a counterclockwise direction. The two systems are close enough that their edges will meet, with the southeast-bound eastern edge of the Biscay system joining forces with and augmenting the similarly southeast-bound western edge of the North Sea system. The two combined streams of southeast-bound air will then pass rapidly across the countryside that lies in between—which is to say, very broadly, France. Recalling, of course, that thanks to the conventions of old, a southeast-bound stream of air is known as a northwesterly wind—for that is what the Mistral is, usually: a northwesterly, cold as the devil and, as its name, *Mistral*, implies, master of all over which it blows.

For once the wind is in place in the skies high above central

France, topography takes over. Both the shape of the land, as well as the Venturi effect—mentioned earlier as being the reason the westerlies in the United States that converge on Mount Washington in New Hampshire become so fast, so cold, and so record-breakingly violent. Here in southern France the wind direction is from the northwest, and the mountain is an outlier of the Alps, a long, bald-topped pastille of limestone called Mont Ventoux. The effect remains the same. The wind howls down from the northwest, rises between Ventoux and the main Alpine chain, accelerates, cools, and settles between the cliffs on either side of the River Rhône, which then acts as its main chute toward the south of France. There are variations on this principal theme, to be sure, but basically it is a cold and disruptive wind that can blow for days on end, and which is an accursed phenomenon for those who have habitually to deal with it. Strabo, the Greek geographer who made the first attempt to fashion a definitive account of those parts of Europe and Africa with which Athens was familiar, wrote scathingly of the Mistral: *an impetuous and terrible wind which displaces rocks, hurls men from their chariots, breaks their limbs, and strips them of their clothes and weapons.* He might have been overdoing it a little; but today most who live and try to play in Provence dread the arrival of the Mistral: it rips tiles from roofs, scatters truffles during a hunt, turns outdoor springtime weddings into amiable shambles of blown tulle and shredded chenille and floods of bridal tears—soon wiped away, because one of the legacies of the Mistral, once it has passed, is the sparklingly clear weather, cloudless skies, fresh air, and general feeling of good health and lively colors that invariably follow in its wake. True, the wind is inconvenient and disobliging when it is blowing—but it nearly always leaves the Provençal people more cheery and carefree than before it arrived, when the air was sultry and languid and full of dust.

Having said which, a mistral is reckoned also to trigger black moods among the local population, most especially when it blows incessantly, noisily, dustily, and irritatingly, rattling the shutters and chilling the bones for days and nights on end. Men in particular can be driven mad by its persistence, it is claimed; and it has long been understood in the courts of southern France that an eruption of domestic anger—even the murder of a spouse—can be blamed on the endless frigid shudderings of a mistral in full spate. Such is not codified in any French law, nor ever has been; but judges and magistrates presiding over a *crime passionnel*—which the Gallic magistracy has in any case long regarded with some sympathy—are generally believed to acknowledge that the blowing of a mistral during a crime's commission is an extenuating circumstance of some merit. *The wind made me do it*, in other words, has some value as a defense.

Most of the old stone farmhouses of Provence and Languedoc have been built with their broad backsides to the north, their entranceways on the sheltered southern side. Church bell towers

The power of a cold winter mistral, a strong and sustained northwesterly that affects much of southern Provence, especially near the city of Marseille. Like the Santa Ana in California, the Mistral is blamed as a mood-changer, and courts in Languedoc and Occitanie are said to display tolerance during mistral outbreaks when dealing with crimes of passion. American judges are less forgiving.

are usually open-sided wrought iron to allow the wind to pass through and leave the tower unscathed. A locally popular Christmastime image for Nativity plays has a shepherd holding his hat down over his ears with both hands, since hats are particularly vulnerable to the wind's unexpected gustiness. Up on Mont Ventoux the shepherds' pathways have snow poles ten feet high, so deep can the drifts be blown; and the summit, when seen from down below, is seldom other than white—either because of actual snow, not unexpected at its altitude of 6,270 feet, or else because its treeless summit is covered by a thick scree of pure white limestone.

Because of the way the houses of Marseille, especially in the older neighborhoods like Le Panier, are so tightly packed, with its inhabitants living hugger-mugger in great urban confusion, they suffer rather less than many a smaller place, like Avignon, which is said by many to be the windiest city in all of Europe. In *Tristram Shandy*, Laurence Sterne goes to some lengths to talk about the wild wind of Avignon and the surrounding country, where the "*windyness*" was such a pronounced cultural feature it has become "*a proverb*"—one of which the people of Provence are, despite the wind's persistent inconvenience, more than a little proud.

{ 5 }

Six hundred miles to the east, however, and to the people of the historically confused Italian city of Trieste, the story of the Bora is much different, and on many levels. For a start, while the Mistral, like most winds in the world, owes its creation largely to a difference of atmospheric pressure, the Bora is manufactured almost entirely as a consequence of differences in *temperature*. The Bora is a katabatic wind, so-called from the Greek word for *downhill* because it involves a mass of air moving rapidly down a slope from up where it is cold, and thus dense and heavy, down to neighboring lowlands where it might become warm and, in consequence, less dense.

As it happens, the strange complexity of Trieste's deeply unpleasant wind is perfectly matched to the strange complexity of city's own recent history. As succinctly as possible: Trieste, a deepwater port on the eastern side of the far north of the Adriatic Sea—Venice is across that stretch of water, a brief ferry ride away—was initially taken by the Austro-Hungarian Empire and provided Austria, hitherto a landlocked country, with its own port and the ability to have both a navy and an eventually very large commercial shipping fleet. But then Italy flexed her muscles, and in the aftermath of the First World War was granted Trieste and the nearby coast as thanks for being part of the victorious Allied powers. The Second World War reversed that situation, however, and with the Italians now the fascist enemy, the city fell in concert with the defeat of Nazi Germany—New Zealand and Yugoslav troops carrying out the actual liberation. Now, though, the postwar world had taken on a profoundly new complexion, and the fine and architecturally rich port found itself pinioned between a defeated Italy to the west and a newborn nation, Yugoslavia, with all the factionalism and religious and historical rivalry of the Balkans included, to the east. Moreover, the new Yugoslavia was essentially within Moscow's orbit, its leadership Marxist, albeit in a peculiarly Balkan way. Trieste was now, and precisely, at the southern terminus of the dividing line that Winston Churchill would later term the Iron Curtain.* Yugoslavia

* His speech on March 5, 1946, at Westminster College in Fulton, Missouri, contained the following famous paragraph, which set the defiant tone for the ensuing half century of the Cold War: "From Stettin in the Baltic to Trieste in the Adriatic, an iron curtain has descended across the Continent. Behind that line lie all the capitals of the ancient states of Central and Eastern Europe. Warsaw, Berlin, Prague, Vienna, Budapest, Belgrade, Bucharest, and Sofia; all these famous cities

was in the Eastern Bloc; Italy was part of the West; and Triestinos, some 330,000 of them, had to choose. Rome, backed by London and Washington, tugged Trieste one way; Belgrade, backed by the Kremlin, the other. No one would budge. But, equally, no rifles were cocked, and no missiles rained.

Instead, a decade of industrial-grade dithering ensued. The United Nations declared the city a Free Territory and oversaw its division into two zones—the more northerly being governed from 1947 by a succession of puzzled senior soldiers: a New Zealand general, two American colonels, and finally, up until late 1954, two British major generals. A secondary zone of rather less utility was run by a succession of Yugoslavian flag officers and diplomats. Then, in 1954, the sovereignty of the more northerly zone was by mutual international agreement handed over to Italy, while the remainder was given to Yugoslavia—a compromise with which everyone, the wildly polyglot Triestinos included, found confusingly satisfying. And then the Cold War ended. Yugoslavia then imploded and dissolved in sympathy with the collapse of the Soviet Union, and Trieste's original northern neighbors of Slovenia and Croatia emerged from the rubble. Today's Trieste, steadfastly overlooked by most despite its nobly imperial buildings, its long and pleasing Adriatic coastline, and the wooded karst hills that rise to its east and south like a protective enfolding arm, rests on its history with a sense of bewildered pride.

And suffers patiently from the knowledge that the hills to the east and south form, in fact, anything but a protective barrier, for there are notches and narrow valleys hacked in places in the great limestone walls, and they allow down into the city the bitter cold

---

and the populations around them lie in what I must call the Soviet sphere, and all are subject, in one form or another, not only to Soviet influence but to a very high and, in some cases, increasing measure of control from Moscow."

winds from the Dinaric Alps that have pummeled and punished Trieste for centuries. These gateways are where the Bora gets in.

The Bora is a cruel wind, bitterly cold, laced with rain or sleet or hail or snow, given to great unpredictable gusts. It tips out of the karst valleys like ice water from a chipped saucer and then swirls into the city with terrific ferocity. It cannonades wildly off the enormous rococo buildings that the Viennese erected all around the administrative center of their one great port, and in doing so it upends vehicles, dislodges streetcars from their rails, pulverizes anemometers, and knocks people off their feet in an unforeseen instant. In fact, for years the city fathers attached thick ropes to the lampposts so that those wishing to cross the street could hang on for dear life when the Bora started to blow. Policemen, their helmets tipped over by the gale, will still help trapped and frightened people as best they can, but often both helped and helper will fall victim to an unanticipated gust and tip

The formidable power of the Bora, with its biting wintertime northeasterly gales, has left an indelible effect on the city of Trieste and its remarkably stoic population. The simple act of crossing the street when the Bora is blowing can be a trial, with the city once providing stout ropes onto which the more frail or frightened might cling.

over in concert, spreadeagled and so looking, appropriately, like an Austrian Imperial heraldic symbol, especially if there is ice on the cobbles, which there often is.

To be sure, Triestinos, proud of their vastly muddled history, are also more than a little proud of their wind. In the center of the city, across from the immense Piazza Unità d'Italia, a long pier juts into the Adriatic. It is the Molo Audace, a berthing quay for the sixty Lloyd Triestino liners and visiting warships that once voyaged to and from this upper Adriatic port. Now there are no ships of consequence, and the pier is, like many a city pier today, a place for the *passeggiata*, with the accompaniment of an accordionist or a juggler or two.

At the pier's westward end, the flagstones glistening with spindrift, stands a waist-high marble column bearing on top a solid bronze wind rose. A Slovenian student was leaning against it, tracing the unfamiliar Latin names of the four cardinal winds of the Mediterranean, the *Greco*, the *Sirocco*, the *Libeccio*, and the northwesterly *Maestro*, linguistically connected to its cousin in Marseille, the Mistral. I told him I was writing a book about wind and had come to Trieste to see—and here he stopped me and gestured down to the wind rose, his finger jabbing between Greco and Maestro. And there, gleaming from the polish of a thousand jabbing fingers, was a full-cheeked cherub, blowing a bora from his full lips—the wind of the city made human, annealed into the public mind on a compass rose of bronze. The student told me he lived up on the bare limestone plateau where the wind is born, and that his mother made spiced pork sausage, known all across the Balkans as *Kranjska klobasa*, and hung it outside her window in the cold whenever the Bora swept by. He jabbed at the cherub. "We love the Bora," he said in halting English. "It is our special wind."

It must be one of the few of the world's winds to have its own museum. Tucked away on a side street, via Belpoggio, the museum is just a single ground floor room in a rather decrepit apartment building. Rino Lombardi, middle-aged, bald, bespectacled, and with an irrepressible enthusiasm for his chosen subject, was an advertising copywriter when, for unexplained reasons, he decided to sell sealed cans of bora wind—allegedly—and advertised them for sale or exchange around the country and the world. To his delighted astonishment, people started sending containers of local winds—allegedly—back to him, from Halifax, Oslo, Mount Fuji, Chicago, Padua, Rio, and Quebec, sealed in bottles, jars, and pots with tags and labels specifying where and when they were each collected. This ever-growing archive of aeolian treasures forms the centerpiece of what has since evolved into a quite serious collection. Much of what is crammed with lumber room insouciance into this tiny space is of little value. But as Joan Sullivan, a Canadian photographer of renewable energy projects based in Quebec, notes, the collection Lombardi has assembled, of *sculptures, kites, weathervanes, anemometers, miniature windmills, wind socks, crampons, pinwheels, whirligigs, flags, hats, maps, poems, postcards, posters, paintings, photographs, cartoons, newspaper clippings, books, documentary films, audio and video recordings . . . illustrates the infinite ways our lives are touched and shaped by the wind: mythical, historical, cultural, political, architectural, meteorological, literary, journalistic, artistic, technological, or just plain whimsical.*

And Signor Lombardi has amassed a fine library, some books popular, some unreadably serious, some academic. In short: the small, undistinguished-looking and open-by-appointment-only room at 9 via Belpoggio is a gem, a precious thing, and must be protected and preserved, its founder commended and honored.

Before we left he made each of us—there were five visitors, all strangers but soon curiously bonded by a common fondness for a museum devoted to the invisible—construct a pinwheel from a sheet of foolscap paper, a pin, a pencil, and a glue stick. I took mine down the hill past those well-marked places where (with a statue) Italo Svevo had lived, where Sir Richard Burton had been

Britain's consul, where (with another statue on a canal bridge) James Joyce had lived and drunk and written, and finally to the pier where the wind rose stood glinting in an evening sun that was now setting over faraway Venice, across the sapphire sea, and on which the Bora's polished cherub blew silently from the northeast.

The pinwheel kept on turning all the while. And on the way, not a single Triestino had looked at me with puzzlement; for to the people of this isolated, melancholy, and fascinating city, an obsessive interest in wind is a perfectly acceptable and natural form of behavior. For my part, wrote Joyce, *I love the bora. It acts on me like a spirit of health. It brings me air from the sky.*

{6}

Wellington, the capital of New Zealand, is by general agreement the windiest city on the planet. The last time I was there a brutal gust blew the spectacles clear off my face, and they were flattened by a passing truck. It is a bustling port at the head of a large, circular bay—its corniche popular with early-morning runners—that lies at the southern tip of the country's North Island. A number of regular ferry services have long connected it to the South Island across the notoriously stormy Cook Strait. The shortest crossing takes three hours and runs between Wellington and Picton. The longest was that which ran to and from the capital along the Pacific coast of the South Island, from the port of Lyttelton, outside the South Island's principal city of Christchurch. The service had been operating for eighty years, and it is fair to say it had become an institution, run by the Union Steam Ship Company, with a modern fleet and a devoted regular clientele.

On Tuesday, April 9, 1968, the northbound overnight express ferry service was being operated by the newest and sleekest vessel of the fleet, the TEV *Wahine*, her initials suitably new and sleek in themselves, standing for Turbo-Electric Vessel. The nine-thousand-ton ship, built on the Clyde and in service in New Zealand for only the previous two years, left the dockside at 8:43 p.m. on what by all accounts was to be a perfectly routine eleven-hour crossing. It would be Good Friday in two days, and so there were passengers going home for Easter. There had been a slight delay because the connecting train from Dunedin arrived late, but it seemed a trivial matter to get into Wellington on schedule at 7:00 a.m. the next morning. She had never thus far been late.

There were 610 passengers and 125 crew legitimately on board, and one stowaway. This was a roll-on/roll-off ship, and on the car deck there were 103 vehicles of varying sizes, from trucks to sedans to motorcycles. The captain was one Hector

New Zealand was proud beyond measure when this sleek and Clyde-built passenger vessel, the Wahine, *unarguably the finest ferry in the world, began service in 1966, working a regular route between the North and South Islands, from Christchurch to Wellington.*

Robertson, fifty-seven years old, an experienced master mariner who knew the ship well and the voyage even better. All was normal in all respects.

Except for one somewhat unusual circumstance: the presence of a powerful tropical cyclone that was heading down on the Pacific side of the North Island. Following the normal naming conventions for South Pacific storms, this one had been named Giselle (her predecessor had been Florence), and she had been born five days earlier as a light circulatory system just to the east of Guadalcanal in the Solomon Islands. She had been traveling south and gathering strength over the warm waters of the Coral Sea while doing so. As with all circulating low-pressure systems south of the equator, the winds were blowing around the storm's eye in a clockwise direction. As *Wahine* left her berth, the weather reports were unsure of exactly where Giselle might be at the time she was due to enter the bay of Wellington Harbor; but *Wahine* was a powerful and well-found ship, and if the seas were high and the winds strong, she ought to have no difficulty. making her way safely to her destination.

Strangely, though, there was little mention at the time of the existence of a second northbound storm, a cold front, heading up from Antarctica. It was possible—but not mentioned—that the two systems might somehow intersect and amplify each other's more dramatic effects. If they were to meet, then they might well do so somewhere near the entrance to Wellington Harbor. Of such a possibility the local Meteorological Office was silent, supposing only that Giselle would probably pass by harmlessly, well to the east.

During the night the seas were high but not uncharacteristically so, and most of those passengers lucky enough to have scored private cabins slept well—until around 5:00 a.m., when they were actually in the Cook Strait, tidal and turbulent and in poor weather notoriously a challenge to even the most well-found ships. By the time the *Wahine* was closing on the cliffs of the North Island she had started to roll violently, and the vessel's lights illuminated great white crests on the waves below. Snow was being blown across the decks by a gathering southerly gale. It was decidedly unpleasant up top, and little better below. One passenger ordered a morning cup of tea, and the steward had the greatest difficult placing it on a table, eventually giving up and ditching it in the sink.

The captain was awakened at 5:45 a.m. and staggered his way to the bridge, where the helmsman and the officer of the watch delivered the first bad news: the radar seemed to have stopped working. The scanner up on the mast was rotating as normal, but on the screen all was gray and indistinct, whatever range was chosen. A vague line on the screen seemed to indicate an approaching coastline; but the lighthouse that marked the entrance to the harbor was nowhere to be seen.

However, the charts were accurate, and the course—more or less due north into and then through the narrow harbor entrance—was already set and the gyrocompasses were all working, so there seemed no good reason to change plans. A radio message from the company said that a tug would be standing by in case *Wahine* needed help getting to her berth. And so, just before six, while it was still pitch-dark outside and the gale was gathering strength by the minute, and with his broken radar rendering all on the bridge nearly blind, Captain Robertson decided to head into the harbor entrance as normal. He took command of his ship. He made the fateful call.

At this point the entrance into Wellington harbor is just about a mile wide—the cliffs of Pencarrow Head to starboard, the peninsula of Seatoun to port, and between them a dangerous cluster of submerged rocks and skerries—marked by an orange-colored light buoy at their southern tip—named Barrett Reef. As the ship bucked and rolled ever more violently—her two propellors were frequently out of the water, so steeply did her bow plunge into the waves ahead—it became steadily more and more difficult for the helmsman to get *Wahine* to respond to his turning of the wheel. Then suddenly, and out of nowhere, came two impossibly ferocious waves and blasts of wind; the first pushed the ship hard over to port, toward the reef, and the second picked up Captain Robertson and flung him clear across the bridge, slamming him against the broken radar console, stunning him. When, after a few seconds, he regained his composure, he decided it was now going to be impossible to get his vessel through the entranceway, and his only prudent course was to turn around and head back out into Cook Strait and the relative safety of the open sea.

But by now neither he nor anyone else on the bridge knew where their ship was. Outside was as black as pitch; there were thundering waves, the winds howled pitilessly, the radar was out, and the vessel was all but unsteerable. The master tried turning by using the two engines as rudders, calling down for the port engine to be set to full ahead and the starboard to a full stop, and then at one stage ordering both engines slammed into reverse to try to halt their wind-pushed northward drift. And still the gale worsened, the cyclone bands now blowing at well over 100 mph, speeds never before experienced in New Zealand, nor by any of the experienced mariners on *Wahine*'s bridge.

And then, the inevitable: The stricken ship, already perilously close to the southern tip of Barrett Reef, was blown relentlessly northward until she smashed into the submerged rocks with a glancing blow to her starboard stern—a collision that tore off her starboard propellor and gashed the side of the hull. Within seconds her port engine died too, and the ship was now utterly helpless—blind, without power, and being battered and smashed by a still worsening storm.

Captain Robertson ordered the two anchors to be released, to stop his ship being blown still farther onto the reef, or onto the cliffs of what was now the lee shore of Seatoun. It took his crewmen fully twenty minutes just to reach the anchor capstans, so vicious were the seas breaking over the foredeck—and then, once the chains had been unpinned and released, the move turned out to have been quite useless, as the ship, blown by the incredible winds like an enormous spinnaker, dragged her anchors along the seabed toward the inevitable.

For now she was taking on water. The captain had indeed closed all watertight doors, but the two car decks, which extended the entire breadth and most of the length of the five-hundred-foot-long vessel, were flooding from a gash in the starboard side. The physical phenomenon of free surface effect water, sloshing from side to side as the ship rolled and thus increasing the angle of the vessel's roll progressively, was what eventually doomed her. Whether Captain Robertson knew this from the moment he learned of the presence of water on the car deck will never be known; he did not announce it but only advised everyone to return to their cabins and retrieve their life jackets and await further orders.

Some passengers later recalled that a curious sense of calm then settled on the ship. Many passengers who at first put on their life jackets soon took them off and used them as pillows while they tried to sleep through the storm. The actual shock of the vessel hitting the rocks had seemed to most quite minor, the sound of the awful crunch masked by the ferocity of the wind and waves. Many supposed they would still be docking by midmorning and the ship would be making her return journey to the Christchurch port of Lyttelton as usual later that same evening.

Nothing could be farther from the truth. Television had only recently come to New Zealand; but dawn had now broken and already TV crews were racing through the gales to be able to see and film—in black-and-white—the extraordinary sight of a huge and ultramodern liner being battered by a storm just off the coastline. Usually ship disasters occurred well out of sight; now a tragedy was taking place in full view of the people of the country's capital city.

In quick succession came a number of dramatic moments.

Crew members on a tug fired a towline that the *Wahine*'s seamen managed to secure to the stern, but as the tug started to heave away an enormous wave threw both vessels upward and the four-inch-thick line snapped. Then, from another tug, an officer managed to leap aboard the wildly pitching deck of the ferry in order to help the master. This newcomer promptly saw what no one had thus far admitted to noticing: *Wahine* was developing a list to starboard. A list was something permanent, a tilt that was developing progressively over and above the terrifying rolling caused by the gale. It was caused by the gathering of hundreds of tons of water down on the car deck, causing a lean to the right that was getting steadily more and more apparent. All on the bridge now knew the awful truth: there would soon come a time when this nine-thousand-ton ship, the pride of New Zealand, would list past a point of no return. Would capsize. Would founder. Would be wrecked. Would sink.

In Wellington Harbor on April 9, 1968, and in full view of the horrified population of New Zealand's capital city, the Wahine, caught and blown onto rocks by a massive gust of cyclonic wind, listed, foundered, and sank with major loss of life. She had been in service for less than two years.

By a little after noon the situation was stark, and Captain Robertson had no alternative but to issue the formal order to abandon ship. Through a crackling and fast-failing public address system, the chief purser made the announcement that no one who had boarded in the starlit calm of Lyttelton just hours before had imagined or contemplated: they were to proceed to their boat deck, to their muster stations, but only on the starboard side. Many had no idea what the word meant. The right side, when facing forward—but down among the endless and now darkening passageways, which way was forward? The low side. Go to the low side.

The ensuing hours were marked by a succession of tragedies. Overcrowded lifeboats swamping, turning over, being run down by other vessels. Cold, shivering, frightened elderly people managing to get to shore but, their craft overturned by the surf, being impaled on the rocks by the beach. The road between Wellington and where many of the lifeboats came ashore being blocked by storm debris, help arriving too late for rescued people dying of exposure in the sand. Children slipping out of mothers' arms and being engulfed by roaring walls of white water, never to be seen again. Policemen in their helmets up to their chests in ice-cold water trying desperately to pull lifeboats ashore against the terrible tug of the undertow. Infants crying with terror and cold. Bodies washing up on beaches. Bewilderment. Shock. A stunned silence. The wind was dropping fast as Giselle moved away down south and the cold front passed by up to the north. It took the waves some time to still, but as they did so, the *Wahine* rolled over for the last time, sank, was declared a total loss. Fifty-three people died. And television recorded it all.

The proximate cause of the sinking of the *Wahine* was the flooding of the vehicle decks. The ultimate cause, however, was the wind—beginning with that single massive southeasterly gust and the two huge waves it triggered, when just before 6:00 a.m. the ship was off Pencarrow Head and trying to make it into the safety of the harbor. The American Meteorological Society's

Glossary of Meteorology defines a *gust* as *a sudden brief increase in the speed of the wind. The duration of a gust is usually less than twenty seconds. It is of a more transient nature than a squall, and is followed by a lull or slackening.* On April 10, 1968, in the ocean approaches to the windiest city on Earth, there was no lull, no slackening. That most pitiless of gusts essentially lasted for the eight following hours, causing what is still remembered as the worst maritime disaster in New Zealand's modern history.

{ 7 }

During the closing months of the Second World War in Asia, more than sixty Japanese cities were near-wholly devastated by American bombing campaigns. We mainly remember today just the two centers, Nagasaki and Hiroshima, that were effectively obliterated by the two atomic bombs dropped in early August 1945, which helped persuade Japan to sue for peace. What tends to be overlooked is the ferocious and pitilessly unremitting bombing campaign that preceded the use of those nuclear weapons and that did every bit as much to bring Japan, already near helplessly on the ropes, to her knees.

This earlier bombing campaign required the use of long-range aircraft, Boeing B-29 Superfortresses, which could operate only from well-established airfields. They could not, for example, operate from carriers. As Japanese forces were slowly driven from eastern China, so a few airfields became available to the planes of what was then still called the USAAF, the United States Army Air Forces. But the distance to Japan was such that only a scattering of cities in the far west of the nation were brought within range, and the raids had little significant effect. Not until islands like Saipan, Tinian, and Iwo Jima—and, ultimately, Okinawa—were purged of the enemy could the B-29s fly across the Japanese homeland with impunity.

All that then remained was a question much debated still

today as being of great moral significance: whether to conduct precision bombing from high altitudes against factories and facilities and so degrade the Japanese war effort, or else to go after the civilian population and try to break the nation's spirit, in preparation for what was then a presumed American ground invasion later in the year.

Civilians turned out to be the victims. Though many, both then and since, would consider it a decision based more on racially inspired vengeance than on operational need, such anti-civilian attacks would indeed become, from the spring of 1945, America's overarching strategy. The pilots were ordered to operate at low altitudes during the nighttime raids and to drop their bombs almost carelessly onto the wood-and-paper homes of the sleeping cities. Moreover, they were to use a newly developed incendiary weapon that would cause immense and appallingly destructive fires—a weapon that, in starting these fires, would also create its own wind, an intense whirlwind of gales that, in fanning the flames, would make the damage even more cruel and widely spread.

The weapon chosen for use over these war-weary and highly flammable cities was a new American invention: napalm.

Sixty-six cities were originally designated as targets. Kobe was the first, as a test run. Then came Tokyo, which was of course primus inter pares. (Kyoto, with its ancient temples and cultural treasures, was omitted from the list.) All the target cities suffered mightily. The casualty figures, beyond those of Hiroshima and Nagasaki, were appalling, yet little noted either then or in the years since. As Robert Neer, a Harvard-educated Columbia professor who wrote a biography of napalm, would later put it, if sardonically, *The atom bombs got the press, but napalm did the work.*

And napalm did the work with the unanticipated and yet vital assistance of its coconspirator, the wind—wind that was born, developed, and magnified by the ignition of the very napalm itself, a self-actuating, self-generating physical phenomenon. The napalm provided the spark, one might thus say; the wind made it into a conflagration.

Napalm is very much the child of Harvard University and a secret wartime program listed on its books as "Anonymous Research No. 4," which was headed by a popular and somewhat Messianic figure, a tenured chemist named Louis Fieser. His group, which specialized in exploding things, was part of a much larger academic effort established by the White House to come up with inventive and original tools with which to prosecute the post–Pearl Harbor war. The Manhattan Project, which devised and built the first atomic weapons, turned out to be the most notorious and costly of these tools. But there were other thinking-outside-of-the-box creations too, paid for by FDR's National Defense Research Committee: bazookas, amphibious landing craft, sonar, radar, proximity fuses—and napalm.

Moreover, Fieser's remit had a very specific link to Pearl Harbor. The Japanese invasions of Malaya, the Dutch East Indies, the Philippines, and Burma suddenly denied the supply of locally harvested rubber to the American military, which had used it as a crucial additive to gasoline in their primitive but militarily very useful flamethrowers. Fieser was asked to come up with a substitute—a gelatinous and highly flammable substance that would burn for a long time and stick to its target while doing so. *Adherence* was the key: gasoline and melted rubber were cruel enough, but the White House wanted something altogether more vicious, and Louis Fieser—a man so revered by his students that even back in the 1940s they had colored T-shirts made with his image emblazoned on them—was set to the task.

Popular scientific lore holds that napalm was born on Valentine's Day 1942. There is just a morsel of truth to this, in that during the winter of that year Fieser's lab was impressed by a mixture stirred up by his team of student helpers—"my 'war boys,'" he called them—of two brown powders, one aluminum naphthenate and the other aluminum palmitate, which, when combined with gasoline and stirred energetically, produced a thick, tarry, and sticky goo, ideal for use on the battlefield. Tests showed that it worked exactly as specified—huge amounts of fire and flame and smoke, great flaming gobs of it thrown around and proving well-nigh impossible to tear away from burning skin and fatally difficult to extinguish. But easy to work with: the gel being simple to pour into torpedo-shaped bombs that were entirely safe to handle until they fell and self-ignited with white phosphorus and TNT bursters that would tear the bombs open just before landing.

A triumphant Fieser needed now only to name his new invention. Accordingly, he took the first two letters of naphthenate and the first four of palmitate and made them into as artificial a new word as Kodak or Kleenex, Hoover or Xerox: napalm. He wrote up his experiment and offered up the name to the NRDC—and he happened to do so on February 14, 1942, hence the Valentine's Day story, which combined love and death in their legendarily perpetual embrace.

Only the supplier of the aluminum palmitate then dropped a small linguistic wrench into the yarn. The substance that he supplied, he admitted sheepishly, was actually made from the oil of a South American babassu palm and had only 8 percent palmitic acid in it, and a whopping 42 percent of something called lauric acid. And so the name of the newly made weapon by pedantic rights should be *nalaur*, not *napalm*. Louis Fieser, euphony on his mind, told the chemists to button their lips. Napalm, right or wrong, was henceforward the name to use, then and ever since.

It became a wildly popular munition right from the start, though at first as a liquid to be fired from flamethrowers. It had the consistency of Jell-O in the soldiers' backpack tanks, turned into a liquid as it was shot under high pressure through the air, then reformed itself into a sticky goo* when it landed on its target,

---

* It behaved *thixotropically*, in other words, acting much as paint does when stirred in the can—almost solid at first, becoming liquidized when agitated, displaying

burning and adhering all the while. Suddenly every force commander in the field wanted some.

Napalm was first used in battle in Sicily in 1943, then in the North African desert, then in the Normandy invasions. It was cheap, infantrymen found it easy to use, and it was lethally effective, by all accounts the most valuable weapon offered to soldiers to help them win on the battlefield.

There was much debate as to how best to deliver it to targets in Japan. Since no American forces were then in the country—though the offshore and western Pacific islands were beginning to fall, one by one, and close combat (in which, of course, flamethrowers were supremely useful) was becoming a leitmotif of this phase of the Pacific war—the planners had to come up with remote forms of delivery. Planes flying in from aircraft carriers were initially the only realistic method. Heavy bombers were an obvious vector, but with bases in western China the closest secure American airfields, most of Japan was out of range. Until islands like Saipan, Guam, and Tinian had fallen and new runways and hangars were built atop them, other means had to be devised. A scheme as outlandishly Pythonesque as any involved, of all things, bats.

The plan hatched back in Washington was to harness tiny napalm bombs fitted with dissolving-chemical timers to thousands—millions, even—of sleeping bats, which could be collected en masse from the rocky roofs of Carlsbad Caverns in New Mexico. At a USAAF base nearby they could be fitted with their lethal carry-ons, put into cages, and shipped across the Pacific, before being placed on small carrier-based aircraft and flown to somewhere convenient (to the bats, who had to fly the final miles) off the Japanese coast and released. Louis Fieser

---

unpredictable resistance when moved in one direction but not in another, thixotropy is molecular-level behavior that has a great capacity to intrigue.

found the idea instantly appealing, performed some back-of-the-envelope arithmetic, and declared that "bomb-delivery by bat was 3.7 times as effective as by gravity" and so was worth trying out.

Experiments were thus ordered. They didn't go too well. Countless unforced errors were made from the very start, and thousands of the tiny mammals died long before their planned immolation, or else they escaped, set fires on barns in neighboring fields, and destroyed an Army Air Forces general's staff limousine. Bats, in short, proved as unreliable as bomb-carrying weapons of war as any rational person might have imagined. But an eventual $24 million of public money was nonetheless spent on a yearlong demonstration of this supposition being true, and in the spring of 1944 the project—Project X-Ray, it had been termed—was finally and deservedly put to painless rest.

In its place came Operation Meetinghouse. This was the brainchild of the newly appointed head of America's strategic bombing campaign, General Curtis LeMay, who realized that the use of his B-29 long-range bombers to drop precision bombs from high altitude in daylight over a generally cloudy Japan was a mission on a road to nowhere—the newly discovered jet stream at that altitude simply blew most of the bombs off target, Japanese air defenses during the day were good enough to knock a few B-29s out of the sky, and Japanese fighters, even at so late a stage in the war, were aggressive and accurate in their attacks on incoming bomber squadrons. In LeMay's view, the only way to hit at Japan and bring her to her knees was to demoralize the civilian population, just as the Royal Air Force had done with its raids over Dresden and Hamburg in the closing days of the European war.

It was the cruelest of strategies, and both the Pentagon and the White House knew it. Years later LeMay remarked drily that had the war against Japan been lost by the Allies, then for deliberately aiming bombs at civilians he would probably have been tried as a war criminal. But the war was not to be lost, and his career and reputation—despite a long-standing and lingering controversy over the specific firebombing of Tokyo—went from strength to strength in the postwar years.

LeMay oversaw the firebombing of no fewer than sixty-seven Japanese cities, once his strategic program was fully underway. After training on mocked-up Japanese villages and towns at the Dugway Proving Ground in Utah, his attacks started with a modest experimental operation over the city of Kobe early in February 1945. That proved sufficiently successful that he ordered the truly massive and horrendous Operation Meetinghouse to take place a month later, during the night of March 9.

Three separate armadas of Superfortresses took part in what would become and remains the most lethal attack made by American forces in any war ever fought—with many more casualties than in any other single city, even including those destroyed by the two atom bombs later that year.*

It was midnight. The pathfinder aircraft came first, descending rapidly until they were barely five thousand feet above the center of the capital. They overflew a ten-square-mile area where it was known that in every one of those square miles lived and worked some 103,000 civilians—the intended targets of LeMay's decisively macabre new plan. The first planes dropped their hundred-pound M47 incendiary bombs, each of them four feet long and eight inches wide, with square tail fins and a detonator at the tip, and painted purple to identify them as holding eighty pounds of napalm gel. The pathfinders dropped their bombs in

---

* The near-total destruction of Tokyo with napalm bombs also proved highly economical to American taxpayers, as statisticians later calculated. After amortizing the costs of the A-bombs' development, it can reasonably be said that Hiroshima and Nagasaki each cost $13.5 billion in 1945 dollars to destroy. The napalm carried by the aircraft that dealt a similar fate to the city of Tokyo in March 1945 cost just $83,000.

two lines that, once they had exploded in walls of yellow flame, were arranged in the form of a cross four miles long and three miles wide—marking for the following aircraft the aiming point on which they should now drop their own bombs. The fact that flaming crosses had other connotations, especially in the American South, was not lost on commentators of the time.

There were some 350 following bombers, joining forces over the Sea of Japan from their bases in Guam, Saipan, and Tinian. As they flew toward Tokyo they could see below the flashes from artillery fire on Iwo Jima, where another decisive attack was taking place at the very same time as their own. The formation was several hundred miles long, and one by one, every few seconds, each plane wheeled down from its cruising altitude to the five-thousand-foot bombing height that LeMay had ordered and released its payload. The carnage that ensued was memorably horrific, and eyewitness stories of what transpired in the gathering fire caused by the explosion of fully one thousand tons of napalm on an area of houses built mainly of cedar wood and paper *shoji* screens can only be imagined.

Most citizens were asleep when the bombers arrived, and they were bewildered and still in their nightclothes when the unquenchable fires started. They poured out into the streets, running for any brick building they could find, running for the bridges, for the rivers, for any body of water that might offer sanctuary or relief. But waters boiled and were no help at all. Napalm gobs were everywhere, on fire, inextinguishable. The temperature rose steadily as flames roared in from all sides, and soon people were dying—quite literally exploding into flames, witnesses said—simply by breathing in superheated air. "Slaughterhouse bombing" the radio called it the next day, and an estimated one hundred thousand people died that night, and a million, maybe even more, were suddenly rendered homeless. In fifteen square miles of central Tokyo, not a single home was left standing, and as dawn broke the rivers were found to be choked with bodies; charred corpses on the street were carbonized and would crumble into mounds of black dust if they were touched by rescue workers trying to clear them away.

It was not the napalm alone that caused the devastation but also the terrifying onset within the blaze of severe surging and self-generating winds—winds that could blow untethered humans for hundreds of yards through the fires they were trying to escape.

Robert Neer, napalm's biographer, quotes the recollection of one Seizo Hashimoto, who was thirteen at the time of the raid. "He saw a woman, dressed in a red kimono with gold and silver threads and a gold *obi* sash, with red lotus blossoms in her hair . . . seized by the [wind] and whipped away and twisted in the air, and ignited: a human torch. A piece of her kimono swirled through the air and dropped at his feet."

When American aircraft dropped napalm-filled incendiary bombs in vast numbers onto the sleeping city of Tokyo in March 1945, a self-generating fire-laden windstorm all but destroyed the entire Japanese capital, causing many more deaths than in the later atomic attacks. Napalm, first made at Harvard, was used against sixty-four further Japanese cities, with civilians invariably the primary target.

These gigantic, hurricane-force gales, it turned out, were generated by the fire itself, which was of such proportions as to manufacture an entirely new meteorological phenomenon that came to be known as a *firestorm*. The existence of such winds had already been observed and recorded at those European cities attacked by incendiaries—Dresden, Hamburg, Kassel, Darmstadt—but no one was certain of the process. The Tokyo raid was different, in that high-altitude American aircraft—the *masters of ceremonies*, they were dubbed—could see exactly what was going on below. The fires created a huge cloud that rose rapidly high above the city just as a thundercloud might in other circumstances; as it rose and rose, sometimes as far as the tropopause and into the lower stratosphere, so it left a low-pressure area directly beneath it, causing fresh air to rush in from all sides at tremendous speed, providing more and more of the oxygen that was urgently needed by the insatiable hunger of the raging fires.

Such a cloud now has a name and a formal weather system classification: it is a *cumulonimbus flammagenitus* cloud, with a code abbreviation *CbFg*. On the ground it causes both major hyperlocal effects—such as the kimono-clad flying woman—and, hypolocally, the super-spreading of the fire itself. Firefighters' efforts to contain such autogenerating conflagrations are invariably paltry and worthless—they certainly were in Tokyo, and they were later reported to be quite ineffectual in Nagoya and Osaka and in the sixty-four other cities similarly attacked by American napalm during the spring and early summer of 1945. And there is no doubt whatsoever that the firefighting efforts were of no value at all in the culminating raids with atomic weapons on Hiroshima and Nagasaki in August.

In both these cases, the fires were so intense that the towering CbFg clouds they each generated were confused by observers with the typical "mushroom clouds" of debris thrown up as the immediate consequence of the earth being ripped asunder by the bombs.

The CbFg clouds came later, rose higher, and were marked by the presence of ice crystals and lightning flashes and whirlwinds and tornadoes and rain bursts and thunderclaps and internal bouts of turbulence and violence. They were, in other words, weather. The fires below had created weather above, which in turn made for the blowing of winds down at ground level that generated yet more fires that were destined to become self-perpetuating, eternal flames.

The fires stopped only when they ran out of fuel—no more houses to burn, no more human beings to immolate. Or, in wider strategic terms, from the American military's point of view, when there was no more napalm available to drop, and no further cities to drop it on.

And also, most important, when at noon on August 15 came the thin and reedy voice of Emperor Hirohito announcing to his people, and to the world, that since "the war situation has developed not necessarily to Japan's advantage" he had ordered his government to—though the word was never used—surrender.

It is often still glibly suggested that the dropping of the atom bombs ended the war. There are several reasons, well beyond the scope of this book, that make it abundantly clear that this is not so, that the use of these bombs had an altogether different purpose, more geostrategic, more to do with the Soviet Union, more to do with revenge, more to do with experimentation, more to do with cruelty and arrogance. What really won the war and brought it to its end was the utter destruction of the sixty-seven cities bombed with Louis Fieser's mixture of aluminum naphthenate and babassu-palm-extracted lauric acid, the misnamed napalm. That, and napalm's cooperative coconspirator, the wind.

It will be remembered from some pages back that in the thirteenth century Japan was spared from Mongol invasion by the unanticipated arrival of a wind that scattered the enemy fleet. The wind was given the name *kamikaze*, the wind of the gods, the divine wind. The same word was then employed again during the last days of the war, describing those brave airmen who dive-bombed their explosive-filled planes into American warships, suicide missions all. One might argue today that the winds generated by the urban fires after napalm attacks—and that turned into ultradestructive firestorms—might, from the Allied point of view, reasonably be described as kamikaze too—divine winds that helped bring to its end a monstrous war.

Or else they were the very opposite: diabolical winds, manufactured to cause maximum suffering and death, and of a cruelty unimaginable to any but the most inhumane.

{8}

"Good afternoon to you. Earlier on today, apparently, a woman rang the BBC and said she heard there was a hurricane on the way. Well, if you're watching, don't worry, there isn't."

This comforting and rather endearing six-second utterance was made on television at lunchtime on Thursday, October 15, 1987. In Britain at least it has since become a cultural trope, so enduring that it appeared on a montage of famous British television moments shown to the world during the opening ceremony of the Olympic Games in London.

It was a weather forecast. The presenter, familiar before, famous since, was an avuncular and vaguely professorial type named Michael Fish, all cardigans and Hush Puppies and mustaches. By all accounts he was a diligent and respected forecaster—the BBC's practice at the time being to employ real meteorologists rather than the merely handsome and plausible.

Except there were two things markedly wrong with what Michael Fish declared to the nation that autumn afternoon. The first relates more to the ethics of broadcasting and need not detain us here: there was no woman calling on the telephone—Mr. Fish by his own admission had simply invented her to add a little human interest and spice to his report.

The second matter is why Mr. Fish's six seconds have achieved such stellar notoriety. *A hurricane on the way . . . don't worry, there isn't.* Well actually, there was. The night of October 15, 1987—and most especially the wee hours of the following Friday morning—provided the occasion for the worst and most destructive windstorm to afflict England for three hundred years. Not since 1703 and the lethally violent series of weather events that killed fifteen thousand people and prompted Daniel Defoe to write a book, *The Storm*, which was almost as well received as his *Robinson Crusoe*, had England been so badly hammered by a wholly unforeseen gathering of winds. The best Mr. Fish could do after his reassuring nostrum was to point to his weather chart showing a low-pressure area of 974 millibars centered ominously over the Bay of Biscay, just a little south of Ushant, the island off far western Brittany, and continue *but having said that, actually, the weather will become very windy, but most of the strong winds, incidentally, will be down over Spain and across into France as well.*

He could not have been more wrong. Just before midnight, while all England slept, the depression hovering over the Western Approaches that had been expected to move eastward inexplicably changed direction and was now seen on radar to be racing *northeastward*, in the direction of London. Moreover, it had deepened from 974 mb down to an unusually low figure of 953 mb, something seldom seen outside the tropics. The storm was undergoing unusually rapid cyclogenesis, with the bottom falling out of the central low-pressure zone and the circulation around it starting to spin wildly and ever more rapidly in the classic counterclockwise direction.

It was at this point that Britain's Met Office in Bracknell, west of London, began to issue a series of gale warnings and forecasts of winds of Force Eleven on the Beaufort scale—*exceptionally high waves as tall as fifty-two feet . . . the sea is covered with long white patches of foam that blow into froth . . . visibility is greatly affected . . . widespread vegetation damage . . . damage to many roofing surfaces.*

But even these warnings turned out to be qualitatively incorrect, treating the inbound storm as a rain event—when, in fact, and incontrovertibly, it was to be an event quite dominated by ultrastrong wind.

Just two institutions, both critical to the nation's safety, received warnings that were accurate enough to be useful: the London Fire Brigade and the Ministry of Defence. The first was told that some solid structures in and around the capital might well collapse under the pressure of the coming gales; and the army was told it might need to be deployed—with ministerial permission—to assist the possibly overstretched civil authorities in the event that this weather system proved catastrophic. But others—most notably the National Grid, which directs electricity from power stations to the population centers—were never warned, nor were the ambulance services, which would have to find and ferry any injured for treatment. *They never asked*, was how a weary weather forecaster remarked when interviewed on television the next day.

And yet still the populace slept. Fitfully, in many cases, but for other reasons: a significant financial crisis was unfolding in the United States, the New York stock markets dropping more violently than for many years. The grim suggestion of a worldwide financial meltdown was abroad. It was widely assumed that London would follow suit—except that from about 2:00 a.m. that night, Britons turned out to have other matters on their minds, to do with the movement of air rather than stocks and shares.

For at around that time the winds began to whistle, then howl, then roar deafeningly. Windows started to rattle. Roof tiles began to slither and crash to the ground. The creak of trees outside intensified, and there came the sounds of cracking branches, and then the more terrible noise of breaking trunks and the crash of whole fallen trees onto cars and roofs, and then of parts of houses being suddenly demolished. Terrified dogs whipped up a chorus of barking. Cats fled and hid. Car alarms were triggered. Windows began to bow inward under the growing pressure of the air, and in some cases they suddenly shattered, covering the floors with broken glass.

Lights went on as panicked people, roused from sleep, tried to work out what was happening to them, to comfort crying children, to make sure elderly parents were secure. And just as suddenly as they went on, so the lights started to flicker, to dim, and then to go out. For good. All over the country, in a ragged line from rural Dorset through Hampshire and Sussex and Kent on up to London, electricity pylons and gantries gave way and toppled, and power stations' cutoff protocols took them offline, slashing power supplies to hundreds of thousands, and then, in due course, to many millions.

It was a long, terrifying, and very dark night, and when dawn finally broke a little after six the next morning, the scale of the devastation became apparent, shocking everyone. England hadn't seen ruin like this since wartime. Back then bomb damage was local; here wind damage stretched for more than a hundred miles. The gales were still blowing hard. Trees were down everywhere. Roads were blocked. An ancient windmill in Kent caught fire, the sails forced by the ferocity of the winds to move against the brake, the friction setting the wood ablaze. Traffic jams for those who ventured out were immense. Huge, well-found ships had been blown up onto beaches and stranded. A liquified-natural-gas tanker—a floating bomb, they warned—had broken her moorings and was thrashing around inside a harbor and being chased by tugs. Trains were halted, aircraft grounded. Buildings were left roofless and broken. Historic piers, built by Victorians for the seaside pleasure of visitors, were washed into the violent seas, crumpled as if made of cardboard and tissue.

The human damage was dreadful. Scores of people were killed, most of them crushed under fallen trees, some drowned in capsized fishing boats. Thousands were injured. Every single London ambulance had been called out—something that hadn't happened since the war. The stock market—which had anticipated high drama and yet more massive losses to echo the unfolding events in New York—had to be closed, mainly because electricity had been cut. Some thought this a blessing.

Fifteen million trees were blown down by the winds—five thousand of them across railway lines, blocking them for days. Famous and notable trees were damaged or destroyed. Six of the seven oak trees that gave the name Sevenoaks to a historic town in Kent were brought down. Well-known and beloved foreign specimens in Kew Gardens and Hyde Park lay flat and wrecked, and the national pine tree sanctuary, home to the world's most definitive conifer collection, lost fully a quarter of its stock.

Not all were wholly dismayed by this involuntary felling, however. Craft carpenters, who for years had applied to parks and arboretums to cut some of their surplus trees but had been re-

London was severely damaged by an unusually powerful extratropical cyclone that came out of the Bay of Biscay on the night of October 15, 1987. It caused death and destruction on a historic scale, but did, however, briefly delay a financial crisis pummeling the rest of the world, since the London Stock Exchange was forced to close.

buffed by officials who regarded all standing trees as sacrosanct, no matter their age or condition, now found themselves awash in timber of the rarest quality, gifted by an act of God. The Duke's Oak in Windsor Great Park, planted in the seventeenth century, had been blown down; the Seven Sisters beech trees in Knole Park in Kent were casualties; a host of ancient sweet chestnut trees in Petworth Park in West Sussex were reduced to scrap in a few blustery seconds; a yew tree, treasured in life for its crowning majesty, in Selborne Park in Hampshire was similarly brought low, as were a cluster of hardwood hornbeams in Epping Forest, in northeast London. All these fine-grained, knot-free, and antique woods were given to, taken by, or sold at craft auctions to men and women who sawed and sliced and mortised and tenoned them into exquisite tables and chairs and decorative pieces that bear the distinction of being given new purpose after the night terrors of the Great Storm of 1987.

In the reckoning that followed, for the meteorologists at their headquarters in Bracknell, who had so egregiously and publicly failed this test, an underlying question niggled away: Had this been a true hurricane, and if so, what was a hurricane—a tropical phenomenon, surely—doing in the temperate latitudes of northwest Europe?

The answer was a classic fudge: the meteorological community declared the Great Storm of 1987 not to be a hurricane at all but an *extra-tropical cyclone*—in other words, a storm that has all the characteristics of a hurricane, to wit: *a counterclockwise rotation of air around an intensely concentrated area of low pressure with the sustained speeds of the rotating air breaching the necessary threshold of seventy-five knots*, but not being fueled by tropical heat of any kind, since its origins were not in the tropics at all. The origins seem to have been, quite simply, within the Bay of Biscay. To this day one wonders—and those who do can only speculate that the Biscayan disturbance that became the Great Storm had earlier origins still—whether it might have been the tail end of a true hurricane that had already done its business in the Caribbean, had then weakened and recurved as spent hurricanes are often wont to do, passed back into the Atlantic proper, and had traveled, almost imperceptibly—certainly undetected—back east under the steering influence of the prevailing westerlies that are found in 49 degrees north latitude, which is where it began to be visible and perceptible. Upon arrival in the often storm-tossed and, as it happened in 1987, unusually warm waters of Biscay, it somehow regenerated itself and turned into the nightmare monster of which we now know and remember so much.

All is speculative history, riven with uncertainty. Cassandras might reasonably say that with global warming keeping such waters warm and making them ever warmer, such storms in England may yet occur more frequently, surely more often than once every three hundred years.

{9}

A true Atlantic hurricane has a generally well-recognized point of conception and thereafter progresses toward its target by way of a generally well-established birth canal. Indeed, any cyclonic storm—whether born in the Atlantic, the Pacific, or the Indian Oceans—has the same mechanics of origin, though not of course the same point of origin, nor passage to where it eventually lands and, after maybe making big trouble, eventually fizzles away. The mechanical process is always the same: A patch of sea in that part of the ocean where cyclones are born is warmed by the sun. The air above this patch of water is warmed too, and rises. Cooler and heavier air from all around the patch rushes in to replace the risen air. The amount of air that rushes in is enormous—and the speeds with which it rushes inward can be, even at this early stage, formidable. But what makes the situation interestingly complex is threefold. First, the air is rushing in from all sides, from all points on the compass. Second, the earth is spinning and, under the influence mentioned before of the Coriolis force, this circle of flowing air is forced to move itself, initially quite gently, in a counterclockwise spiral—or, south of the equator, clockwise. Third, the upwelling warm air that prompted all this surface movement cools as it moves higher into the atmosphere, and if it is humid air—as most likely it is, since it is rising from the sea surface—it loses its moisture as rain, and this seemingly simple act of becoming rain releases all the energy that placed the rain-to-be in the cloud in the first place. This free energy has to make itself useful somehow, and invariably physics chooses to do so by pulling the now rather drier air still farther upward and spinning the now developing vortex below it ever more rapidly. A giant engine, in other words, has just been cranked into life; fueled by the heating of the sun and the condensation of the rain, and driven by prevailing winds and turned by Coriolis's legendary force, it expands outward from its hollow, calm, low-pressure center and makes its way toward its intended target.

A tropical thundershower off the West African coast in the bay now known as the Bight of Benin seems to be where the majority of Atlantic hurricanes like this are born; and if, by maritime and climatological and meteorological good fortune, this otherwise insignificant squall happens to be nudged westward by a lurking trade wind, and if in the nudging it happens also to be persuaded to rotate, counterclockwise, and if that spinning disturbance survives its passage until it is clear of the African coast and manages to get itself out into the open ocean, then after this, great geostationary satellites will look down and eventually spot it and then track it as it roils and ripples its way through the atmosphere—like watching a rabbit running through a tall wheat field, making ripples as it goes—until it becomes big and powerful enough to do damage, by which time it has been given a name by various of the world's weather services, and then aircraft are dispatched to fly over it and peer into its most intimate recesses and measure and compute its various statistics and prod it and poke it and worry about it, and its existence makes the television news and the name is on everybody's lips, maybe for only a few days or, if it performs truly magisterial mayhem, maybe for longer, and possibly then it enters the history books—as with Katrina, for example, a devastating event that was once a small seasonal thundershower in the Bight of Benin but went rogue, off the rails, off the charts, became delinquent, and will be remembered, at least by all Americans, now and for many years to come.

{ 10 }

In the late autumn of 2004 I was given what most journalists would consider a dream assignment, by a well-known American monthly magazine. The commissioning editor knew I was interested in weather—he knew that I kept a weather station in my garden, had a sunshine recorder on the lawn, and had a barograph inside the house and that I reset it every Sunday, thus having twenty years' worth—more than a thousand weekly graphs—of records showing the rise and fall of the atmospheric pressure in western Massachusetts sitting in a library cupboard. He thought, based on all this, that I might care to track the progress of a hurricane as it formed, developed, and made landfall, and have the piece ready for publication ten months hence, for the magazine's following August issue. *Weren't hurricanes born somewhere in West Africa*? he inquired. Well, go there, and then to all the other places that help tell the story. Eight thousands words, due in June.

I spent the winter planning. Reading, poring over maps and weather charts—most preciously the Admiralty Pilot Charts of the North Atlantic Ocean, one for each month of the year and with the ocean divided into ten-degree squares with an elegant wind rose in the center of each, displaying the prevailing winds' direction and strength, a cartographic and meteorologic creation like few others and a thing to treasure, whether researching hurricanes or not.

I arranged such visas as I thought I might need: I had the idea, before I knew about the squalls in the Bight of Benin, that a good starting point would be the site of a small desert whirlwind in some West African capital, somewhere like Ouagadougou or Nouakchott, Abidjan or Niamey, or maybe even from a coffee shop in Cotonou, which, being in the Republic of Benin and on the sea, actually abuts the Bight and had among its population figures—fishermen, weather forecasters, scientists—who would know a great deal about the formation of storms locally and so might offer me a colorful way to begin my essay.

I would then contrive to spend time on the waterless African islands of Cape Verde, with their surrounding seas frequently playing host to the true beginnings of westbound hurricanes. At this stage and in this part of the world, the feature would still not be noted for being particularly violent, nor even dramatic; just a warm-water depression, probably still deserving only of a number and not yet a name, but with a circulating structure more or less intact, a pattern both observable and recognizable. It passes by Cape Verde, and only the weather forecasters at the Hurricane Center in Florida and the US Navy's Fleet Weather Center in Norfolk, Virginia, take notice, diligent through night and day. If the ocean waters to the west of Cape Verde are suitably warm, then the depression will be fed with the kind of fuel that it needs to become an entity of note, after which Miami and Norfolk sit up and begin to take serious notice and the danger signals begin to sound; the citizens of the landmasses that are next on the depression's path, the Lesser Antilles and the Windward Islands—islands like Antigua, Dominica, St. Lucia, Guadeloupe, Martinique. St. Vincent, Grenada—turn on their radios and begin to listen for updates on the track of the slowly turning body as it whirls steadily toward them.

And all this planning of mine—visas for Benin and Mali, names of friendly sorts in Praia, contact numbers for navy forecasters in Virginia and their civilian counterparts in Miami, and then the names of hotels in the Caribbean, and later for Cuba, Jamaica, Key West, Mobile, the Louisiana coast—all was being painstakingly contrived during the bleak winter months of December and January amid the snow and ice storms of western Massachusetts. It all sounded like great fun, utterly fascinating, and too good to be true.

As indeed it turned out to be. I had booked my flight and bought my tickets for the first leg, a direct flight from Newark to Lagos in Nigeria, from where I would begin my West African explorations. Then the call came in: it was early February, and on the phone was the previous autumn's commissioning editor; he was, to say the very least, embarrassed.

The magazine's executive editor had looked at the budget for the proposed story and decided it was going to be too costly, and had canceled the assignment. I was to stand down, unpack my bags, and curse the wind. Metaphorically, of course. The only blessing, the chagrined caller said, was that at least I didn't take the call in a hotel in downtown Ouagadougou.

That was February. On August 23, Hurricane Katrina, the costliest cyclone ever to hit the United States, totally devastated large areas of New Orleans, Louisiana, killing nearly 1,400 people, destroying thousands of houses, and causing an estimated $125 billion in damage, being among the most savage hurricanes in American history. The catastrophe was poorly handled by the government and had devastating political, social, and cultural consequences.

And the cruel irony was not lost on anyone at the magazine, which just ten months before had been so eager to put the story of the birth of a hurricane on its late summer front cover. Had it managed to do so, everyone with even the slightest interest in this formidable wind-driven disaster in and around New Orleans would have turned to our story, in the one journal that would have had the picture down pat, of how and why this had all happened. In the event, the cover story was a speculative essay about fiscal mismanagement, which nobody read at all. The executive editor left the magazine shortly thereafter.

Katrina was indeed a monster storm. Yet in this particular case, the strength and speed of the wind was just one of the factors that caused it to be so calamitous. There were two chief villains. The storm surge was one; the waters of the Gulf of Mexico were pushed by the wind and, to a lesser extent, hoisted upward by the extraordinarily low pressure of the storm's eye, and waters twenty to twenty-five feet higher than normal broke through the levees that had been built to protect the lower-lying areas of New Orleans proper and caused sudden flooding on a biblical scale.

{ 11 }

The mention of levees serves as a reminder of the second villain of the piece: the fact that the early city fathers ever allowed New Orleans to be sited where it was, at the mouth of a capricious river, right beside the sea, on land that was below sea level and in an area generally known for violent storms. The hubris of their having done so is a common component of early American settlement more generally: cities built in waterless deserts, above active seismic fault lines, close to volcanoes, and in places where ill winds and flooding is certain; and, in the latter case, where enormous and costly constructions—seawalls, anti-ocean barriers, levees—have to be raised to keep the city intact and its inhabitants safe. Wind-generated storm surges also devastated the Texan coastal city of Galveston back in 1900—once again a low-slung city protected by an equally low-slung seawall, with victims—the poor here, as most often—being sucked out to sea in their hundreds, drowning in an oceanic maelstrom they could never have imagined invading the presumed safety of their city.

Eight thousand people are believed to have died in the Galveston hurricane of September 8, 1900, which to this day remains the deadliest natural disaster in American history. Though its West African point of origin is speculative, something meteorologically notable was first recorded as a somewhat organized proto-cyclone on August 27, twelve days before landfall, by a cargo ship passing about a thousand miles east of the Windward Islands. From there its track was, by today's standards, fairly typical—and it might have been fairly accurately forecast when it passed over Cuba on its way northwestward had not old enmities, born of the Spanish–American War, reared their heads. On orders from President McKinley's White House, the chiefs at the National Weather Bureau in Washington forbade any use of the word *hurricane*—as Cuban forecasters had done in their descriptions of the coming storm, since to do so was to give the weathermen on a former Spanish territory more credibility than, in Washington's view, they deserved. So the Texan civil authorities were simply warned of inclement weather, took little heed—and the sea rushed in and dragged thousands out to drown. But the sea, not the wind.

There still have been, throughout recent American history, any number of wind-forward catastrophes, as distinct from disasters that can be classified—if the distinction may be permitted—as water-forward or fire-forward, and that place the London windstorm of 1987 in a proper context. Preeminent among the named storms are Hurricanes Andrew, Camille, Charley, Harvey, Hazel, Irma, Michael, Rita, Wilma, and then in 2024, Milton, and they each jostle for pole position depending on the three usual factors that make the news: the highest sustained wind speeds (for one uninterrupted minute) and thus their assigned category according to the dictates of the Saffir-Simpson Hurricane Wind

Scale;* the pressure in the hurricane's eye, the lower the number equating to the more severe the storm; and the estimated cost of restoring the affected area back to normal, a figure usually supplied by the insurance companies that will have to foot the bill.

Television has stripped high-energy weather phenomena of much connection with reality. Given the various stations' and their parent networks' endless search for ratings, forecasters have too often become performers in dramas of their own creation—storms are tracked, forecasts for ever more dire situations are offered, colorful maps and charts present terrifyingly lurid displays of the likely outcomes, reporters wearing ever more garish items of foul-weather gear are filmed being buffeted by gales or drenched by rainstorms, their fates entertaining and alarming viewers in equal measure. Invariably such entertainers oversell the storms that, on reflection, seem to have deserved a more sober recounting. And comparisons are invariably made, as they are with sporting histories and statistics. So one might hear that 2024's Milton had the highest-ever sustained wind speed (180 mph, recorded over the spectacularly warm—87°F—waters of the Gulf of Mexico, before its landfall in western Florida), that Harvey in 2017 had the most rainfall (sixty inches over Houston), that Wilma in 2005 still has the record for the most intense

* Herbert Saffir, the creator in 1969 of the scale that bears his name—along with that of the meteorologist and then US National Hurricane Center director Robert Simpson—nursed a lifelong passion for making public structures as strong and safe as humanly possible. His obsession—he was a civil engineer—may well have been born from his traumatic experience aboard the 1934 burning and wreck of the Havana-to-New-York passenger steamer SS *Morro Castle*, in which 137 died—the causes of the fire and subsequent calamities brought about by poor safety codes. The Saffir-Simpson Hurricane Wind Scale relates likely hurricane damage to five increasingly high wind speeds, with pressure today to add a sixth category for storms with sustained wind speeds in excess of 157 mph. The scale, though useful shorthand on broadcast news, is regarded as inadequate, since it takes only limited account of factors such as storm surge, eye-wall height, and atmospheric pressure within the eye.

storm (with an eye pressure as low as 882 mb), and once again that Milton outranks all other storms in terms of its cost, estimated by some at $246 billion, at least twice as costly as Katrina.

A hurricane makes a suitably vivid appearance in the 1948 John Huston movie *Key Largo,* in which Edward G. Robinson plays a gang boss based in nearby Cuba who holes up in a near-empty off-season hotel in the Florida Keys, planning some complicated kind of heist. As the storm nears, Humphrey Bogart, along with his then wife, Lauren Bacall, ensure a just and fitting conclusion to a classic film noir that holds less appeal for students of the genre and rather more for the amateur weather-obsessed, who can indulge in ninety minutes of meteorological drama based on a real Florida Keys event that occurred thirteen years before—the so-called Labor Day Hurricane of 1935. Memorable events from that storm are drily recounted during the movie by the hotel's owner, played by Lionel Barrymore—recalling in particular when disaster overcame Henry Flagler's ill-starred Overseas Railroad.

Flagler, one of the original boosters of Florida as a holiday destination for winter-weary New Yorkers, had built, at the turn of the century and at enormous expense and with the use of formidable amounts of manpower, a single-track, 130-mile-long railway line between Miami and Key West itself, with iron and concrete bridges linking island to island to island. His plan—which he saw as an extension of his network of highly successful railway lines that dominated travel all across the Atlantic coast of Florida—was twofold: to exploit the opening of the Panama Canal and provide regular freight service up to the American mainland; and to offer regular passenger service from Key West to Havana, back when Cuba, just an hour's ferry ride away, was America's principal Caribbean playground. Neither was a consummate success. The line lost money, and though beloved locally as a means of getting from island to island and up to Miami for shopping expeditions, it never really took off.

It had endured three hurricanes during its years of construction; it would not survive the hammering it took from what has come to be known as the Labor Day Hurricane of 1935. This was indeed a Category Five windstorm; and though it did create a powerful storm surge, the wind was what really flattened buildings by the score, drove people into the ocean and tore down bridges, inundated railroad lines, and, in one celebrated incident memorialized by Lionel Barrymore's character, threw an entire train—except for the engine—clean off its rails and into the mangrove swamp.

In the scene, a clearly terrified and sweating Edward G. Robinson is pacing back and forth while the hotel creaks under the strain of the howling wind. Glasses keep falling from shelves and shattering on the floor. The lights flicker on and off, and there seems a real danger that the entire wooden structure could collapse at any moment. Barrymore tells him that the "worst storm we had was in '35"—and one of the gangsters confirms that he had read about it in the papers—and says eight hundred people were swept off Matecumbe Key and into the sea. Robinson angrily denounces him as a liar and cries that no one would come if that were true.

But laconically, and with a voice immediately recognizable as Barrymore's, the hotel owner continues: "A relief train was dispatched from Miami—the barometer was down to about twenty-six inches [892 mb] when that train pulled in to Homestead. Engineer backed his train of empty coaches into the danger zone, and the hurricane knocked those coaches right off the track. Two hundred miles an hour that wind blew. A tidal wave twelve feet high went right across the key. Whole towns were wiped out. Miles and miles of track were ripped up and washed away. Nothing was left. More than five hundred bodies were recovered after the storm, and for months afterwards corpses were found in the mangrove swamps."

More than a few movies that depict violent weather are notorious for their hyperbole. This scene, however, has long been regarded as reasonable and accurate; the Broadway play of the same name on which it was based, by Maxwell Anderson, is seen as having a background that was scrupulously researched as well.

{ 12 }

It would be idle to bring up too often the clichéd proverb about *an ill wind blowing nobody any good.* Yet in Hong Kong in the mid-1990s an inbound typhoon proved to a fault how true such a saying can be. It is a slightly unusual little story, personally memorable, wholly trivial to most. I include it mainly to illustrate the British colonial authorities' onetime attitudes, at least on the Chinese coast, to wind and weather.

(And here it should be pointed out that *typhoon* is simply a hurricane by the name used in and around the northwest Pacific Ocean, while *cyclone* is the preferred term in the south and west Pacific and across the Indian Ocean. All three terms essentially describe the same phenomenon: a cyclone—to repeat what was noted here earlier—being a circular movement of air—rotating counterclockwise in the Northern Hemisphere, clockwise south of the equator—that is both well-defined and has sustained wind speeds in excess of 75 mph. All three weather patterns are technically *cyclonic,* from the Greek κύκλος—whirling in a circle—but in Chinese-dominated waters they are typhoons, from *big wind,* and in the Atlantic and eastern Pacific hurricanes, from the Carib *Huracan.*)

First, some background: In Hong Kong the government agency that for the last century and a quarter has been charged with informing the local population of the approach of dangerously inclement weather is the Hong Kong Observatory—formerly, from 1912 until 1997, and when I was living there during colonial times, the Royal Observatory. In the early days of British colonial rule, the warning of an approaching typhoon was made by the firing across the harbor of a typhoon cannon, together with the hoisting on a signal mast in the most southerly peninsula of Kowloon a large black drumlike object, unmistakable in shape and large enough to be visible by moored vessels several miles away.

Signal Hill, now just a tiny oasis of parkland peace hidden among the public housing skyscrapers of southern Kowloon, used to be home to what became a complex array of masts and towers and yardarms and aerials from where a variety of code signals could be displayed. There was also, as in most other port cities around the world, a large black sphere, a time ball, which would be lowered from the top of its mast to the base at noon precisely, so that all masters and commanders of the vessels in the harbor could set their chronometers accurately. As clocks and watches became more reliable, the need for a time ball slowly vanished; but the use in Hong Kong of masts and coded signal flags and shaped objects to convey weather information lingered for decades.

The typhoon gun itself—sounded after a telegraphed warning had been received, usually from the Philippines, four hundred miles south and on the usual track of northbound cyclones—was abandoned as being too difficult to hear, and instead early in the twentieth century especially loud bombs were exploded in the harbor instead, to ensure that everyone was paying attention to the possibility of impending danger. An elaborate system was then devised for the raising of enormous black objects of markedly different shapes and sizes that would be hoisted up high to tell of storms of different kinds and intensities bearing down across the South China Sea, with Hong Kong in their crosshairs. It was a system that worked well for many colonial-era years: from the standby signal represented by the bombs, and then by way of ten numbered signals that were hoisted, from Number One—which warned that a storm was about five hundred miles away and heading toward somewhere on the Chinese coast—right up to Number Ten, which meant that within the next twelve hours a full-on super-typhoon would hit Hong Kong squarely between the eyes. The millions of residents—whether English- or Cantonese-speaking made no difference, since these were audible and visible signals only—were kept aware and, generally, safe.

A vestigial version of the old system continues to this day, with the local radio stations announcing that this or that "signal" having been been "hoisted," even though flags and giant cloth symbols and the hoisting of anything went the way of the dodo soon after the end of the Second World War. The government did its best to enforce the suggestions implicit in the Royal Observatory's warning. Signal Number One—no guns or bombs involved, just a radio announcer's voice, in English or Cantonese—whereupon you were supposed to be fully aware that meteorologic danger might be at hand.

When Signal Number Three was hoisted, city dwellers had to shift movable objects—flowerpots, pets' feeding bowls, plastic chairs, and the like—from balconies, do last-minute errands to stock up on essential supplies, and generally prepare for the worst. And if the worst was to come—something that would be announced by the hoisting of Typhoon Signal Number Eight—then all official work in the colony would cease, government offices would close, Kai Tak Airport would stop all landings and departures, and while city buses and subway trains would work to get people home, those services too would halt, if and when the weather deteriorated further. Hong Kong buses are double-deckers, which tend to get blown flat in tremendous winds.

And here is how an ill wind of this very kind turned out to work a small local miracle.

At the time I ran a small technology company—what today one would call a start-up. We were a year old, had twenty employees, and worked out of an office in eastern Hong Kong, a place called Causeway Bay. It would be fair to say we had no real idea how to run a company—the technology, now long supplanted and hopelessly out-of-date, was what fascinated us—and we were entirely unaware that for some reason we had not been paying the office rent for months.

In Hong Kong, small matters like rent arrears can spiral out of control very quickly—and the first that we knew anything was wrong was when we read in the local paper's High Court Daily Report that our company was being sued, and a petition for our eviction would be heard the following morning. The plaintiff—our landlord—sought to have a judge confirm that we would be barred from entering the office. As it happened, we had at the time precious little money in the till and would not be able to settle the arrears until the day after our eviction. In short: if the court opened for business the next morning and the judge quite reasonably ruled for the landlord against us, we would be tossed unceremoniously out onto the street, and—with word getting around Hong Kong's business community like a grass fire—our company would be doomed.

Except late that afternoon Signal One had been hoisted. A fast-moving cyclonic storm had been spotted in the South China Sea, and it appeared to be heading more or less our way. Maybe it would hit the coast by the Chinese towns of Shanwei or Shantou, about thirty miles to our east—not a direct hit but bad enough, if the storm was especially vicious, to prompt the hoisting of ever more alarming signals.

Just before midnight RTHK, Radio Television Hong Kong, announced the hoisting of Signal Three. Back then you could dial in to the observatory and receive a faxed copy of the latest position of the storm. And sad to say from our selfish point of view, it seemed to be curving eastward, toward the old treaty port of Amoy, now called Xiamen, and thus likely to miss Hong Kong, or merely to subject the place to some short-tempered cold northerly winds, not enough to close the city down. The courts would open as usual; we would be adjudicated as being in default; the fledgling company would die.

I rose at five to see the storm track still edging farther and farther away. It was a deeply depressing moment, and ignobly, I still prayed for us to be struck by a calamitous typhoon, bringing much suffering of millions just to avoid one stab of forgettable commercial pain for this one insignificant company.

And then at six, the clock radio came on. The time pips from Greenwich. The background sound of some hastily shuffled papers. The announcer cleared his throat: *Good morning, Hong Kong,* he said. *And before the news we are instructed to report that the Royal Observatory has just hoisted Signal Number Eight. I repeat, Signal Number Eight.* And then he allowed himself a moment of levity: *No need to go to work today. Everything will be shut.*

I could hardly believe it—especially since I could see, as the dawn began to spread over the hill where I lived, that it wasn't even raining. But the observatory had spoken. The signal had been hoisted. Number Eight. The colony was shutting down for the day.

And that, of course, included the High Court. There would be no hearing, no ruling, and our tiny firm had its death sentence suspended. I went to the court the next day, in my breast pocket a crisp Hong Kong & Shanghai Bank cashier's check for the sum in arrears, plus enough for another three months by way of an apology. When our case was announced I rose and told the court I was sorry and, if allowed to approach the bench, could I hand him the check in settlement? The judge graciously said that he was minded to agree, the plaintiff concurred with a wintry smile, the case was dropped—and the firm survived for many more years, with our board rapping me over the knuckles and making me give assurances that I would determine never to let such a mishap occur again.

We had been saved by the wind.

The weather on that day before, when everything had been shut down, had turned out blustery and cold and with occasional late-morning bouts of heavy rain—but no worse than Manchester on a summer Sunday. Certainly not typhoon weather. Not a drama worthy of hoisting a Signal higher than Three; certainly not Number Eight. The chart showed this clearly: this wretched little storm had headed up toward Taiwan, missing us by many miles.

The Royal Observatory had made a mistake.

The papers were apoplectic, and such politicians as were allowed to speak out against the colonial authorities raged. There soon came an explanation of sorts. Six weeks before there had been a truly dangerous typhoon that did indeed hit the territory directly on the nose, but in that case the observatory had been caught flat-footed and did not issue a warning in time. On this occasion they decided to overcompensate under the rubric of better safe than sorry and started warning that any approaching storm might well hit us directly and at full strength—and so on this day, *our* day, decided to close everything down out of what is nowadays called an abundance of caution. Hence the grumbling: Thousands of shops and restaurants did no business, the stock exchange was closed down for a few hours, and the lifeblood of Hong Kong, its flow of money and ability to make more of it, was briefly halted.

There were no other serious casualties in either event. And we, to our surprised delight, survived. I took the staff out to dinner the day that our case was dropped. There were two toasts: the first to the Royal Observatory, and the other to Typhoon Signal Number Eight—a lucky number in all possible senses.

{ 13 }

Severe Tropical Cyclone Tracy, very small and tightly wound and an altogether different weather system in an altogether different corner of the Pacific world, came ashore early on Christmas morning in 1974, and its immense winds blew an entire city down so comprehensively that, for a while at least, no one in the entire country seemed to know where it was, or what fate had befallen it.

The city was Darwin, capital of Australia's Northern Territory, the country's "top end," and on the hot and humid midsummer night of Christmas Eve, its forty-seven thousand inhabitants had no idea of what was about to happen. It was expected to be Christmas as usual: twinkling lights on the palm trees, artificial snow, gifts to wrap, carols on the local ABC radio station, midnight mass, peace and goodwill to all men. If the air was thick with heat and the rains beat down in bursts—well, this was the season all knew as The Wet, and as the local pun had it, it came with the Territory.

Besides, the locals were a hardy lot, born of a rude necessity. Those who made their living in this young and faraway tropical outpost were still very much thought of, sometimes scorned, as pioneers, men and women of a very different stripe from those who lived in the softer, more sophisticated and more comfortable cities of the south, like Melbourne, Sydney, and Brisbane, and even those in edgier cities like Perth and Adelaide. These were working Australians—bushies, diggers, battlers, larrikins, mates—who had come north to make a go of it in the harsh climate on the edge of the equally harsh outback. The old-timers who remained had endured nights of Japanese bombing raids during the war, and a sturdy resilience was in their DNA.

Theirs was an isolated life. A single road, the Stuart Highway, connected Darwin with Alice Springs a thousand miles away in the country's red center. There was a short, narrow-gauge railway line mainly used for the export of iron ore. The air service was fitful at best, and during The Wet the frequent torrential rains rendered such service as existed even more unreliable and limited. And yet, although very few of the more metropolitan Australians ever ventured north to visit the city, there was a countrywide feeling of pride and concern directed at the simple existence of the place and its people: Darwin specifically, and the Northern Territory more generally, seemed to embody everything about the Australian spirit—or the white Australian spirit, it should perhaps be said—that had led to the young country's signal success in the world. Which is probably why the country rallied so very impressively when news of what had happened that terrible Christmas morning became widely and shockingly known.

Tracy, the storm that caused it all, seemed almost insignificant from the moment of its birth until five days later when it attained its full-fledged and murderous maturity. It was first spotted by the American weather satellite ESSA-8—a quarter-ton RCA-made device that could take detailed photographs of the atmosphere from its orbit eight hundred miles above the Earth's surface—early on Friday morning, December 20. It was seen first as a small swirl of cloud in the Arafura Sea, more than two hundred miles north of Darwin. The ground station in California, as a courtesy, informed the Darwin Weather Bureau of activity in its area of interest. But no need for alarms or excursions; this was cyclone season in the tropical waters between Australia and Timor and New Guinea, and this was not even a developed cyclone. Not yet.

Later in the day another American satellite, NOAA-4, which had been up in the sky only for a month, took an infrared image of the clouds and was able to determine that pressure was falling in the pattern's center and that a clockwise rotation was starting to develop. Accordingly, the depression was noted as being a proto-cyclone, and once it became visible on Darwin's long-range radar screens on the Sunday morning of December 22, it was given its name: Tracy. The Americans were duly thanked; Australia would monitor the storm from now on.

As the hours passed, so Tracy steadily deepened, and at the same time, and unusually, became smaller and smaller, like a tightening sphincter. It became ever more compact and concentrated, eventually sporting a diameter of less than ten miles, compared to the often five-*hundred*-mile width of a decent-sized typhoon in the North Pacific, in places like Hong Kong. All of this quite fascinated the duty weathermen tracking her but did not alarm anyone, since Tracy was heading in a straight line southwestward and would pass many miles north of Darwin on her way.

By dawn on December 23, the storm had shrunk to a mere seven miles wide, with its outer wind bands blowing at 40 mph and increasing steadily. Still no reason for action—no emergency warnings, no need to alert the local radio stations and have the sirens start their yowling. Some locally thought the atmosphere a little strange, though. The air was hot and unusually thick—"jerky," as one Chinese shopkeeper later described it. The cloud bank to the west had an ugly, lurid aspect to it, with vivid greens and purples and a thick velvet-black cap overhead. There were glints of faraway lightning and the occasional thunderclap. The local indigenous people, the Larrakia, who lived in the tall grasslands outside Darwin, spoke later of the songbirds having fallen strangely silent, and that the large green ants, normally everywhere at this time of year, had vanished underground. A woman named Ida Bishop, who ran a fleet of shrimp boats, said that "something dreadful" was about to happen.

That midnight, at the start of Christmas Eve, something unusual occurred: Tracy suddenly and inexplicably changed course and started to head directly southward. It was by now well off to the west of Darwin, so still no panic, but interesting nonetheless. And the wind speed kept on rising: 60 mph at dawn, 75 mph at 9:00 a.m.

Until at noon something totally unanticipated took place. Tracy accelerated still further and at the same time changed course once again—this time heading southeastward, directly toward an unsuspecting Darwin. The Weather Bureau put out a Flash Cyclone Warning telling the population that this new and deadly storm would reach the outskirts of the city just after midnight. All of Christmas Day would, in other words, now be dominated by the progress of what was clearly going to be, by all accounts, a most terribly destructive storm.

The diary notes of the Weather Bureau's duty officers record in staccato the events as they unfolded on Christmas Day:

0100: Wind gusts in excess of 100 km/h commenced in Darwin.

0300: Numerous reports of severe damage in and around Darwin. Communications with the mass media were lost when both radio stations failed.

0305: A peak gust of 217 km/h recorded at Darwin Airport.

0310: The anemometer recording system failed.

0350: A thirty-five minute period of calm commenced at the airport.

0400: The Bureau's radar at Darwin Airport showed that the eye was overhead. Rainfall recorded at the airport since midnight totaled 144.2 mm. Atmospheric pressure reading in the eye was 950 mb corrected to mean sea level.

0425: The calm ended at the airport. Strong winds resumed from the southwest.

0430: Radar tracking ceased.

0600: Tracy's centre was located near Howard Springs.

0630: Winds were abating in Darwin.

1100: The cyclone was weakening as it passed the . . . village at Middle Point. After crossing the Adelaide River about midday on the 25th Tracy degenerated rapidly into a rain depression and then moved slowly southeast across southern Arnhem Land into the Gulf Country of Queensland.

By then Tracy had quite ruined Darwin and had left most buildings—many of them insubstantial and quickly built houses up on stilts to keep them above the floods of The Wet—smashed and flattened, and the city's people were left without anything—no power, no water, no food, no shelter, no communication. Forty-seven thousand people—awakened to the howling winds and the screeching, splintering cracking sounds of their homes collapsing wall by wall around them, with the clanging and slicing sounds of thousands of sheets of corrugated metal roofing flying through the night air and doing who knows what damage to whatever each might hit, and with vivid arcs of blue flame when a downed power wire would short-circuit and maybe kill someone who had been touched by it—were suddenly pitched out onto the rain-soaked streets in total darkness, terrified beyond belief by a living nightmare that kept on getting worse and worse. The air was filled with screams from people who would receive no help, whose plight was quite unknown beyond the limits of their frontier town, and who had no ability to summon aid or sustenance or any kind of relief.

Down in the rest of Australia, life progressed as on any normal Christmas Day: wakeful and excited children playing under the tree, gifts being handed out, the living room a sea of wrapping paper, then maybe church, the agreeable if overabundant lunch, silence for the Queen's Speech on ABC, a walk in the park, Christmas cake and carols for tea, a classic television movie, then a nightcap and a deep and satisfied sleep. And if some in Sydney or Melbourne had tried to telephone a relative up in Darwin and the circuit was busy, all was understandable, a typical overstressed line, maybe try again in the morning.

Except that by teatime in Canberra, where the government skeleton crews were still manning the key offices of state, word was somehow getting out—a shortwave radio message sent from Darwin to the ABC radio station down in Mount Isa in central Queensland by a handful of ham radio operators in Darwin who had batteries and were able to bounce their shortwave signals around the world—that Darwin had been devastated by a windstorm, and that nothing, just nothing, appeared to be remaining.

Rescue flights began to arrive at the wrecked and littered airport late on Christmas night; they found a pitch-dark city with a cowed, hushed, homeless, and hopeless population—no panic, no anger, just a grim realization that their city was wholly gone. Christmas was long to be remembered all across Australia as *the day that Darwin died*, and those who were wandering bewil-

Almost the entire northern Australian city of Darwin was destroyed by a massive storm, Cyclone Tracy, beginning on Christmas Eve 1974. The national government, stunned by the scale of the damage, ordered the evacuation of some thirty-five thousand residents, and the city was essentially closed for rebuilding for the next five years. It is now thriving once again, and cautiously said to be stormproof.

dered around the ruins of what most rescuers compared in its appearance to Hiroshima assumed it would never recover, would be thought of as a failed and terrible bad dream. Ninety percent of its houses had collapsed. Of the forty-seven thousand inhabitants, forty-one thousand had nowhere to live. Dogs, terrified and unfed, roamed streets that were no more than ragged pathways through piles of splintered wreckage. All seemed utterly without a future.

But Darwin did indeed survive, and has today been wholly rebuilt. Its entire population was evacuated in the largest civilian airlift Australia has ever known, and the cleanup of the depopulated city began within days in a massive operation about which, to this day, the entire country rightly feels a considerable sense of pride. All the buildings in the new Darwin are said to be cyclone-proof; and in the half century since Tracy, the violent weather systems that with harsh regularity visit Australia's top end—most notable among them Trixie in 1975 and Ingrid in 2005—have spent themselves without major loss of life. And indeed, Tracy killed only seventy-one people—her brevity, her compact size, and her swift passage through ensured that it was the city that died, not its people.

And now the people are back—the same diggers and mates and jawbreaking larrikin pioneers as before, only four times as many of them: 172,000 in 2024. The Larrakia still prefer to live in the tall grasslands outside town; the green ants are back, and the songbirds are once more in full throat. For how long this will be occasioned depends on the wind patterns in the ocean to their north—more generally, the storms in mid-ocean are getting ever more wild, and though there is much talk of the diminishment of average continental wind speeds around the world as part of the so-called Great Stilling, in such places as Darwin, where truly violent storms are an all-too-familiar commonplace, there is widespread apprehension still: Can we yet weather the worst that the wind might throw at us, when it returns, more severe than before?

{ 14 }

Famously, at least in American naval circles, the pivotal role of the violent winds of the tropical cyclone in the future of the world has emerged as significantly more potent than it might once have seemed. This is largely due to an interview given in March 2013 to the *Boston Globe* newspaper by the then commander of 330,000 US forces in the Indo-Pacific region, the four-star admiral Samuel J. Locklear III. In the paper he declared that in spite of the wide belief that China and North Korea posed the greatest security threat to the region, and to humanity more generally, the greater potential foe was worse and wilder weather brought about by climate change. Rising sea levels, ever more destructive windstorms, shifts in populations—all these things, the admiral declared, had the potential to "cripple the security environment" and would, in the long term, prove a much greater existential threat to humankind's future than all the high-decibel geopolitical wranglings that appeared to dominate the news cycle.

His remarks, welcomed by many, were also said to be incautious, and his career—stellar to that point—then stalled, with right-wing American climate-change deniers accusing him of publishing unnecessarily alarmist views. But the facts tended to bear him out, most especially his concern over the impact of the very most dangerous and deadly super-typhoons, of which he said "we are already [in 2013] at number twenty-seven or twenty-eight here in the western Pacific The average is about seventeen."

And right on cue, the very next super-typhoon turned out to be the fiercest—the record still stood in 2024—and most lethal tropical windstorm in recorded history. Admiral Locklear had uttered his jeremiad in March of 2013. On the first day of November, a Friday, the duty navy officers at the Joint Typhoon Warning Center at the admiral's own Pearl Harbor base in Hawaii spotted an ominous cloud pattern developing to the southeast of the tiny island of Pohnpei in Micronesia in the far western Pacific. The island was already well-known to weathermen for being among the world's rainiest places, with three hundred inches falling each average year. But on this occasion, and ever since, Pohnpei has become known as the birthplace of Typhoon Haiyan. This storm did not head south, as Tracy had forty years before; instead, it kept to its latitude of about seven degrees north of the equator and headed due west. Eight days later, with wind gusts measuring 196 mph, it slammed headlong into the eastern Philippines near Leyte Gulf, where it wrought the most terrible damage.

Here there is a point that needs to be made. The casualty figures from Typhoon Haiyan are formidable: 6,500 killed, 27,000 injured, whole cities in the eastern Philippines flattened. The US Navy's prediction was spot-on, and the $21 million rescue mission Operation Damayan, with hundreds of US Marines from their base in Okinawa and the enormous strike group of the nuclear-powered carrier USS *George Washington*, is seen as textbook-perfect, a classic example of soft-power projection. But it is regrettably undeniable that it is Cyclone Tracy, the far less murderous storm, that remains the far better known and more remembered wind-driven tragedy. Tracy resulted in the deaths of nearly seventy people, and yet is known by millions more than know anything at all about the destructive consequences of the winds from Typhoon Haiyan, which killed a hundred times more.

And we all know why this is. The same cynical tenet of Western journalism obtains, which holds that simply appending the two words "in India" to a headline about, say, a major train crash will demote it from being a page-one story to its burial on an inside page devoted to the more perfunctory coverage of foreign news. The city of Darwin was populated largely by white people; those who had lived and loved and worked and idled contentedly on and around the Leyte Gulf in the Philippines were not. Their fate at the hands of a monstrous wind was deemed to have been of secondary interest to the total ruin of a Western community.

The wind, of course, makes no judgment in these matters; it blows equally for all.

And the strength, size, and lethality of Haiyan underscores Admiral Locklear's point: that Pacific cyclones really do seem to be getting larger, more robust, and, when they hit land, more dangerous. The embattled Philippines, for example, was hit by no fewer than four typhons *at the same time* in the late cyclone season of 2024—at much the same time as the United States experienced two Category Five storms lashing Florida and the Gulf of Mexico—a two-for-one event never known before.

The trend, if indeed it turns out to be such a thing, got formally underway in late 1979 with the creation and unprecedented three-week-long-lived typhoon named Tip—which had a low-pressure eye of 870 mb (the lowest on record), windspeeds of 190 mph* (approaching the highest on record), and a diameter of 1,380 miles (by far the biggest ever known, and vastly larger than the near-microscopic-seeming Tracy, which was at its moment of deadliest impact just seven miles from side to side).

In the half century since Tip sprawled itself across the western Pacific, a pattern of sorts appears to have emerged. While any consistent increase in the numbers of cyclones each year is still a matter of dispute—the 2013 cyclone season that so excited Admiral Locklear was a dramatic one, but there have been quieter years in between—the intensity of such storms as have occurred, and thus their wind speeds, has gone through the roof. The UN agency that monitors long-term changes in wind patterns reports a 12 to 15 percent increase in intensity since then, with sub-900 mb low-pressure eyes, once a rarity, becoming an all-too-common phenomenon. And linked to this is the increased rapidity of such intensification—a depression with a central pressure in the low 900 mbs can deepen these days at a speed unimaginable in the middle of the last century. The causative villain in both cases—lower pressures and a greater hurry to get there—appears to be directly related to the steep rise in the ambient temperature of

---

* The winds from Typhoon Tip caused a particularly bizarre tragedy at a US Marine training depot in central Japan. Enlisted men flown in from Okinawa were assigned to sleep in Quonset huts by the training fields. Up on a hill above them were rubber bladders containing thousands of gallons of gasoline. In the middle of the night an especially powerful gust broke the retaining wall below one of the bladders and at the same time split the bladder itself, causing a torrent of gasoline to flow down the hill and, eventually, into the sleeping marines' quarters. Before anyone noticed, the river of fuel hit the huts' space heaters and ignited—causing an inferno that killed thirteen of the marines and injured scores more.

the ocean's waters. Meteorologists are, however, a cautious crew, and there is still no formally accepted link between higher cyclonic wind speed and anthropogenic climate change. Just a suspicion. And voices in the Pentagon asserting that the kinds of winds and waves that their warships are having to sail through these days are getting ever more challenging, and common sense suggests there can surely be only one reason why.

{ 15 }

Whether writers are considering this new phenomenon or one as old as time, literature has long been fascinated by the invisible power of the wind. Two thousand years ago at the very least, the oldest books in the Bible contain numberless references to high winds, often used symbolically to describe God's power, his judgment, or a divine intervention.

Most famously there is the parting of the Red Sea, allowing the Jews to continue their escape from Egyptian slavery: it is described in Exodus 14:21: "Then Moses stretched out his hand over the sea, and the Lord drove the sea back by a strong east wind all night and made the sea dry land, and the waters were divided." Elijah expressed the naked power of the Almighty, as in 1 Kings 19:11: "And he said, 'Go out and stand on the mount before the Lord.' And behold, the Lord passed by, and a great and strong wind tore the mountains and broke in pieces the rocks before the Lord, but the Lord was not in the wind." And to Christians, the all-compassionate Jesus of course always possessed the power to intervene and calm any windstorm that threatened those for whom he cared, as in the Gospel of Mark 4:37 and 4:39: "And a great windstorm arose, and the waves were breaking into the boat, so that the boat was already filling. . . . And he awoke and rebuked the wind and said to the sea, 'Peace! Be still!' And the wind ceased, and there was a great calm."

Shakespeare made much of great winds. *The Tempest* sets the tone with the stage direction at the play's very start: "On a ship at sea: a tempestuous noise of thunder and lightning heard." And then with *King Lear*, the raging storm has for centuries tested the ingenuity of legions of directors: "*Blow, winds, and crack your cheeks! Rage! Blow! / You cataracts and hurricanoes, spout / Till you have drench'd our steeples, drown'd the cocks! / You sulphurous and thought-executing fires, / Vaunt-couriers to oak-cleaving thunderbolts, / Singe my white head! And thou, all-shaking thunder, / Smite flat the thick rotundity o' the world! / Crack nature's moulds, all germens spill at once / That make ingrateful man!*"

{ 16 }

For Daniel Defoe, 1703 was by no stretch of the imagination a good year. He was rising forty and had not yet achieved the fame and wealth that his writing of *Robinson Crusoe*, sixteen years later, would bring him. He had made some money as a businessman, selling stockings and glassware and a perfume made from the glandular secretions of civet cats, which he bred. But he incurred enormous debts and eventually declared bankruptcy, turning then to writing—pamphleteering, initially—in the hope of becoming a well-known figure of sympathy on the London social scene. This he did, but at the price of being arrested and tried for sedition (one of his early pamphlets poked too much fun at a tribe of humorless clerics), and he was sent first to the pillory (where admiring crowds threw flowers at him rather than the customary fusillades of rotting vegetables), fined the equivalent of $100,000 in today's money, and chained up in the infamous Newgate Prison. An admirer paid his bond and had him freed—just in time to experience in its entirety the infamous Great Storm of 1703, which broke less than two weeks after he walked out through the prison gates.

Wisely, he made much of his involvement with the hurricane-force winds of that November weekend to turn himself into a first-class reporter, changing almost overnight from a reckless pamphleteer into, once he had honed his literary skills, a journalist of the first order. A year after the storm engulfed London and the English southeast, Daniel Defoe delivered and had published his first serious book of nonfiction, *The Storm*, which sold well in the eighteenth century and is still in print today.

Here, then, is Defoe, sixteen years away from *Robinson Crusoe*, showing off his literary chops by reporting the very beginning of the Great Storm of 1703 as it broke on the night of Friday, November 26:

> It did not blow so hard until 12:00 at night but that most families went to bed though many of them not without some concern at the terrible wind which then blew. But about one or at least by two o'clock 'tis supposed, few people that were capable of any sense of danger were so Hardy as to lie in bed. And the fury of the tempest increased to such a degree that as the editor of this account being in London, and conversing with the people the next days understood most people expected the fall of their houses.
>
> And yet, in this general apprehension nobody durst quit their tottering habitations; for whatever the danger was within doors, it was worse without. The bricks, tiles and stones from the tops of the houses flew with such force and so thick in the streets that no one thought fit to venture out though their houses that near demolished within.
>
> The author of this relation was in a well-built brick house in the skirts of the city and a stack of chimneys falling in upon the next house gave the house such a shock that they thought it was just coming down upon their heads: but opening the door to attempt an escape into the garden the danger was so apparent that they all thought fit to surrender to the disposal of Almighty Providence and expect their graves in the ruins of the house, rather than to meet most certain destruction in the open garden. For unless they could have gone above 200 yards from any building there had been no security; for the force

> of the wind blew the tiles point blank; though their weight inclines them downward and in several very broad streets we saw the windows broken by the flying tile sherds from the other side; and where there was room for them to fly the author of this has seen tiles blown from my house above 30 or 40 yards and stuck from 5 to 8 inches into the solid earth. Pieces of timber, iron and sheets of lead have from higher buildings been blown much farther as in the particulars hereafter will appear.

{ 17 }

Writers who deal with storms at sea are maybe rather less circumspect, especially when the tempest is at its height. Joseph Conrad might have had a lifelong career in the merchant marine, having spent years on cargo ships bent to every imaginable task, from cabin boy to captain, but he abandoned the sailor's life when he was just thirty-six, partly because of poor health but mostly because he became fascinated by the art of writing, and in English—a particularly interesting intellectual challenge, since he was a Pole—his given name was Józef Teodor Konrad Korzeniowski—and English was very much *not* his first language. His memories of the sea and the countries he visited and the curious ne'er-do-wells and other characters he met are everywhere in his writing, and his descriptions of extreme weather are nowhere better illustrated than in *Typhoon*, the saga of his time in a Siamese-registered tramp, the *Nan-Shan*. We are somewhere in the South China Sea, and the storm is raging:

> The *Nan-Shan* was being looted by the storm with a senseless, destructive fury: trysails torn out of the extra gaskets, double lashed awning blown away, bridge swept clean, weather-cloths burst, rails twisted, light screens smashed and two of the boats had gone already. They had gone unheard and unseen, melting, as it were, in the shock and smother of the way of the wave. It was only later when upon the

white flash of another high sea hurling itself amidships, Jukes had a vision of two pairs of davits leaping black and empty out of the solid blackness, with one overhauled fall flying and an iron-bound block capering in the air, that he became aware of what had happened within about three yards of his back.

He poked his head forward, groping for the ear of his commander. His lips touched it—big, fleshy, very wet. He cried in an agitated tone: "Our boats are going now, Sir!"

And again he heard that voice, forced and ringing feebly, but with the penetrating effect of quietness in the enormous discord of noises, as if sent out from some remote spot of peace beyond the black wastes of the gale, again he heard a man's voice—the frail and indomitable sound that can be made to carry an Infinity of thought, resolution and purpose that shall be pronouncing confident words on the last day, when heavens fall, and justice is done—again he heard it and he was crying to him as if from very far—"All right."

He thought he had not managed to make himself understood. "Our boats—I say boats—the boats, Sir! Two gone!"

The same voice within a foot of him and yet so remote yelled sensibly "Can't be helped."

Captain McWhirr had never turned his face, but Jukes caught some more words on the wind: "What can—expect—when hammering through—such—Bound to leave—something behind—stands to reason."

Watchfully Jukes listened for more. No more came. This was all Captain McWhirr had to say, and Jukes could picture to himself rather than see the broad squat back before him. An impenetrable obscurity pressed down upon the ghostly glimmers of the sea. A dull conviction seized upon Jukes that there was nothing to be done.

If the steering-gear did not give way, if the immense volumes of water did not burst the deck in nor smash one of the hatches, if the engines did not give up, if way could be kept on the ship against this terrific wind, and she did not bury herself in one of these awful seas, of whose white crests alone, topping high above her bow, he could now and then get a sickening glimpse—then there was a chance of her coming out of it. Something within him seemed to turn over, bringing uppermost the feeling that the *Nan-Shan* was lost.

"She's done for," he said to himself, with a surprising mental agitation as though he had discovered an unexpected meaning in this thought. One of these things was bound to happen. Nothing could be prevented now and nothing could be remedied. The men on board did not count and the ship could not last. This weather was too impossible.

{ 18 }

For my money, Richard Hughes remains the master of violent sea storm writing, as the brief excerpt below will, I trust, illustrate.

Richard Arthur Warren Hughes, comfortably born and admirably connected, educated at Charterhouse and graduating from Oxford in 1922, made a name for himself in literary circles while still a teenager, having a review published in the *Spectator* when he was only seventeen. He has the distinction of writing the very first broadcast radio drama that went out over the air in January 1924. His novel *A High Wind in Jamaica*, published in 1929, was an instant success, telling the story of a group of young and winsome English schoolchildren who were accidentally trapped below decks on a late-nineteenth-century pirate ship then scavenging the waters of the Caribbean for booty. The story is cleverly counterintuitive, with the youngsters proving themselves far less morally bound than their villainous shipmates, and was translated into a highly saturated sixties movie, in CinemaScope and DeLuxe Color, that transfixed the more prim audiences on both sides of the Atlantic.*

This book—with its evidently carefully researched nautical

* One of the schoolchildren was played with a confident aplomb that belied his youth by the eight-year-old writer-to-be Martin Amis.

theme—won the attention of a prominent Liverpool shipowner, Leonard Holt, whose family owned a modest fleet of cargo steamers. His literary interest was piqued because in 1932 one of his vessels had become spectacularly trapped, hopelessly pinioned for five days and nights inside a Caribbean hurricane. The ship, the SS *Phemius*, had been near-fatally damaged—not least by having her entire main funnel blown off and tossed into the raging sea by winds that exerted an overpressure of at least two hundred tons per square inch; a normal ship's funnel is designed to resist pressure of maybe fifteen tons, so this mighty wind was quite unprecedented in maritime history, and it was long afterward regarded as a miracle that the *Phemius* survived her ordeal. Holt commandeered his captain's log from the hurricane, declared on reading it that it was a story for the ages, and approached the then poet laureate John Masefield, known for his love of the sea, to write a fictional account. Masefield had no interest but passed Holt on to Richard Hughes, who was wholly intrigued, and after some negotiation agreed to take the project on.

Accordingly the SS *Archimedes* was born, and under the command of a Welshman, Captain Edwardes, she set off from Norfolk, Virginia, with a mixed cargo of dry goods and bound, by way of Panama, for ports in the Far East. She dropped the pilot at Cape Henry at the end of the Chesapeake Bay and set out for San Salvador and points south, before heading with her British officers and largely Chinese crew southwestward toward the port city of Colón, the Canal, and the broad reaches of the Pacific Ocean.

The *Archimedes* would never reach her destination. Five days out from Norfolk, after she had passed between Cuba and Haiti, a stiff northeasterly wind began to blow. It would blow more and more strongly and ferociously and brutally, and would not ease up for the better part of a week, blowing the soon powerless, waterless, unmaneuverable, and half-sinking wreck of a ship before her for hundreds of miles through mountainous seas, terrifying all on board and reducing them to starving wretches within an inch of their lives for day after terrifying day.

But as winds have a habit of doing, the hurricane eventually eased. And since we might feel we have heard a sufficiency of wind-borne death and destruction from Conrad and Defoe and Shakespeare and their kin, here is Richard Hughes picking up the story of the battered and broken *Archimedes* once the gales have dropped and left their awful consequences. This was Monday morning. The storm had taken place on the previous Wednesday.

> Dick Watchett [the ship's most junior officer] certainly had no memory of getting out of his bunk that morning. The first thing he could remember was when he was on deck. It was a limpid and lovely morning, the sea was smooth except for a slight very long, very rapid swell that passed almost faster than the eye could follow it and gave the ship next to no time to rise and fall.
>
> The sky was the blue of a field of gentians, the air clear as glass but warm. The very sea seemed washed, it sparkled so blue, so diamond bright. The blue wood-smoke from the improvised donkey-funnel floated up into the still air and hung there, the only cloud there was, scenting all the horrible litter of the decks with its sweet smell. It was such a morning that you could hardly believe no larks would presently rise ascending on their clear voices into the clear sky. The voices of the woodmen in the donkey-room rose sharp but still faint and the occasional blow of an axe,
>
> Dick heard an order given in a confident voice. There was a hiss as steam-cocks were turned on, then the sudden clanging of the pumps, loud at first until the water began to rise; then steady and slow. They were pumping out number six hold. A brown and filthy stream, creamy with air bubbles, began to cascade into the clean sea.
>
> The pool of brown in the clear blue spread. Presently Dick noticed a queer thing: fish rising to the surface of it, floating dead, their white bellies up. It was so impregnated with tobacco juice it poisoned

any fish who came near it. Imagine all that nicotine flowing through delicate gills!

The pumps could not work for long at the time. The highest pressure of steam the wood furnace could raise was 40 pounds, roughly the pressure of a motor tyre, and they couldn't hold it for long. A brief spell of work and then a rest while they stoked the furnace once more. Meanwhile the brown stain and the sea faded to a yellow opaqueness but the poisoned fish remained floating round the *Archimedes* in the hundreds, with staring eyes and fixed and gaping mouths.

Presently the pumps began their painful and poisonous vomiting once more. It may have killed the fish but he put wonderful new heart into the crew of the *Archimedes* and as the level of water in the after holds fell they sang, they worked like blazes and in their zeal they smashed for firewood even objects that were not really seriously damaged. There were a few of these after all saved in a miraculous way—the bookcase in the smoking room for instance, a flimsy affair with a glass front, it had fallen on its face on the floor and in some unaccountable way not even the glass was broken. Yet a saloon table, I told you, had been snapped off its clamped legs. It was not as if the bookcases contained a Bible—you could not even find a superstitious reason for it being saved, and it only contained ordinary literature.

Another pretty miraculous thing when you come to think of it was that nobody had been killed. Things had been happening all round them as lethal as an air raid, and yet there were no casualties—not even a broken bone. Everyone, nearly, was cut and bruised, but that was all. The worst sufferer was Mr. Soutar: at one moment the heaviest midshipman had been flung onto a particularly bad bunion he had, and he'd yelled with agony he limped from it still.

By sight of a star at dawn and a solar sight later, Captain Edwardes was at last able to fix his position. Being so far from his estimated position the calculation took him some time, and when he plotted the result on the chart he rubbed his eyes. He was away 100 miles north of Cape Gracias: all banks passed. The storm had carried him nearly 400 miles from the point at which it struck, in five days. Moreover it had probably not taken him directly there—curving they had probably drifted at least 100 miles a day, an average speed of four knots, travelled for the most part broadside on. Of course the speed through the water of four knots broadside on was hardly possible—the storm must have carried the sea along with it too, and indeed when he examined the chart he saw that his earlier surmise must have been true, that the sea was raised up near the center of the storm in a flattish cone with a circular motion, only slower like that of the wind, and so they had passed safely over banks they could never have crossed if the sea had been at its normal level.

The first thing he did of course when he found his position was to announce it to the *Patricia*, and when he got her reply he was thankful, for this steam raised on wood, it was after all only make believe. It enabled them to do a bit of pumping, or when in tow perhaps it would work the steering gear. It could work the fans, but he knew very well the fans alone could never get the main furnaces going from cold, without a main funnel he could never really enable them to raise main steam again . . .

{ 19 }

Fierce wind plays a powerfully symbolic role in countless more works of literature, even if it is less central to the narrative arc as in these particular works. Steinbeck's *The Grapes of Wrath* forms the epigraph to this chapter to help illustrate the point; then, eight thousand miles east and a century earlier, the stormy turbulence of the love affair between Heathcliffe and Catherine in *Wuthering Heights* is lashed into sharp focus by the wild gales sweeping the Yorkshire moors; *Moby-Dick* is of course hunted through any number of violent windstorms, but unlike the marlin in Hemingway's *The Old Man and the Sea*, the white whale does not succumb to the wiles of man, and Ahab is the presumed victim; Margaret

Mitchell's evident regret that the ways of the Old South will, in the wake of Lincoln, soon be *Gone with the Wind* is palpable, and excites controversy still; Edgar Allan Poe used foul winds to create the feel of terror and menace; Shelley wrote an *Ode to the West Wind* in which the gale comes to symbolize both destruction and rebirth; Homer and Milton used the violent movement of air on both land and sea to illustrate the nature of chaos; and Katherine Mansfield tackled the ferocious winds of Wellington, New Zealand—already told of in an earlier account here—in a haunting short story of 1915, "The Wind Blows."*

{ 20 }

And then there is L. Frank Baum, an upstate New Yorker who cut his teeth running a store and editing a newspaper in the frontier land of late-nineteenth-century Aberdeen, South Dakota. In a fit of imaginative magic he created a character with a wind-scented name, Dorothy Gale, and had her and her house whirled up into the prairie sky by a notorious storm he called a *cyclone.* Dorothy turned out to be on her way to meet *The Wonderful Wizard of Oz*, would in time become Judy Garland, and would let her terrier, Toto, famously know that wherever they were, they were no longer in Kansas anymore.

The wind that blew the child and her pet to commence what would become their world-famous adventure—a wind that we today know to be the leitmotif wind of the American plains—was of a kind L. Frank Baum would have experienced on all too many

* Not to be confused with *When the Wind Blows* by Raymond Briggs, a 1983 comic novel about the breeze-borne spread of nuclear fallout after an atomic war and how it affects one elderly couple. The book, also made into a movie and a radio play, remains popular in the British Isles.

summertime occasions in and around the cornfields of Aberdeen. Though he called it a cyclone—it was the title of his first chapter—it was in fact something else: It was a type of *tornado,* and it was well chosen on all manner of levels. For although tornadoes do technically descend from the angriest of storm clouds in other parts of the world, they do so far more regularly, and so iconically, within that thousand-mile flatland that unrolls between the Rocky and the Appalachian Mountains. The tornado is, as a consequence, unquestionably and unchallengeably America's wind.

And moreover, a tornado is a wind that, without breaking a sweat, can pick up a house, a child, and a pet and much, much more, and then move on as it will and where it will. Yes, it is America's wind. And a wind with a symbolism all its own, being also clumsy, unstoppable, all-powerful, and with the potential to do terrible damage.

{ 21 }

A tornado is most usually the offspring of a particularly ferocious thunderstorm. It is generally born out of a certain magnificent type of cumulonimbus cloud that some have over the years prettily designated *C. mammatus*—that is, by having pendant mammary protuberances dangling weirdly from its lower surfaces. The creation of these *mammata*—which, if they reach down to solid earth and become discolored by the dust and debris they suck into themselves and so blacken and take on a truly ominous aspect, are newmade tornadoes—is an external manifestation of the utter mayhem that is going on within the apparently stately majesty of a fully-fledged thundercloud: the torrents of hot and humid air rocketing upward, the ice crystals and hail clouds raging where the warm uprush slows its climb and cools, the seeming serenity of the anvil shape where the jet stream has caught the cloud's upper edge and flung it windward, the overfalls of chilled air hurtling back down toward the ground, the lightning triggered by the static charges generated by the phenomenal mechanical shearing as the masses of air pass one another so violently, the peals of thunder, the vortices and shudderings and rotational funfair rides that are anything but fun to those unlucky enough to be caught within a cumulonimbus cloud.

And people do get caught. Recreational fliers—gliders most often—who soar too close in the hope of catching a thermal—there are many recorded cases—can be sucked into such a nightmare of cloud-white mischief. They will be ingested from the safety of their clear air through what must seem like a science fiction event horizon, and then twisted and turned with insane forces as they head uncontrollably upward, their wings and rudders and stabilizers and finally fuselages shredded, leaving their all-too-vulnerable bodies to the grasp of unimaginable and invariably lethal forces. And what such unfortunates are experiencing are not tornadoes as such: they are in the factories where tornadoes are made, experiencing some of the forces that are then distilled into these uniquely American windstorms. And such unfortunates seldom survive; of a group of five Germans who were sucked into a cumulonimbus cloud during a contest in the western Alps in 1938, four died, the horribly injured survivor telling of rampaging forces beyond imagination turning the inside of the cloud into a shrieking vortex of hell.

It is the shear forces that produce the tornadoes. Within the multilayered complexities of the cloud, within all those zones of differing and varying pressures and temperatures, pockets of rain and hail and snow and electrical discharges, there will inevitably be zones where air is streaming by rapidly in one direction and then, just a few feet or scores of feet above it within the cloud, another layer of air will be flowing by in quite the opposite direction, and the two layers will rub against one another with gigantic power and cause forces of torsion and tearing—and shear—that will more often than not translate into *rotation*. It is as though a tube of mixed air is created out of nothing, and this cylinder begins to spin horizontally. Within the already disturbed interweaving skeins of currents and eddies and whirls and whorls, this feature has a kind of elegance, of symmetry, of order—a moisture-rich cylinder twirling inside the crazed disorder at the upper edge of the cloud.

Then one further known and definable mechanical event occurs. The hot-air updraft, a constant and essential feature of the growth and evolution of a cumulonimbus cloud, manages to take hold of one end of the cylinder of rotating air and tips it from the horizontal to the vertical. It happens quite suddenly: the cylinder is lying flush with the lower surface of the cloud base, and then, hoisted by an up-current, it is tipped, one end up, the other down and so poking through the base.

At this point it will become visible to observers on the ground. It will become pendulous, udder-like, hernia-like, a *mammatus*. It will be grey, little different from the cloud base from which it is protruding. And it will retain its angular momentum: it will still be spinning, usually in the direction that cyclones do, but doing so from within a storm that itself is not spinning, for a cumulonimbus thundercloud may internally present a crazed confusion of movement but it is manifestly not rotating itself. Its pendulous offspring, however, is—and if one of these tipped-down cylinders is to become a true tornado, it will nudge itself downward, teasing the onlookers and potential victims, dipping a little down toward the all-too-vulnerable barns and mobile homes and fields of cattle below, then maybe retreating up into the cloud base, then returning, nudging ever closer to the ground.

Once it reaches solid earth, the fledgling tornado begins to scour the land surface, and in doing so picks up soil and dirt and debris, its color darkening all the while, changing from gray to deep green to purple and then to black. The pressure inside the tube—the spinning flinging air molecules ever outward—drops precipitously, becoming a near-vacuum, and with its outer wind speeds accelerating equally wildly until the water droplets are hurtling at speeds that reach into the hundreds of miles an hour, it is vastly faster than any mere hurricane. A tornado requires a whole new scale of measurement—the Enhanced Fujita Scale—which takes over from where Beaufort and Saffir-Simpson left off, and yet does not so much measure wind speed—for the accurate measurement of a tornado's rotational speed is something that is still nigh impossible to achieve—as physical damage caused by the wind. So, an EF2 tornado will totally rip the bark from a tree; an EF4 tornado might destroy a brick-built supermarket; and an EF5 will not just destroy it, it will sweep away all the bricks from which it had been built and obliterate it, utterly. It will be as though it never existed, such is the cleansing power of an EF5 tornado.

A typical tornado—a kind of windstorm almost uniquely American—dangling tantalizingly from the underside of a cumulonimbus storm cloud over Kansas. When its powerfully rotating column does touch down, it invariably wreaks fierce damage, churning across the landscape, cutting large swathes through any city it encounters.

The damage that can be done by such a phenomenon is cruel and capricious. The pressure differentials caused as it passes by can explode houses, shatter glass, deflate human lungs; its wind power can lift fully laden eighteen-wheelers like thistledown and carry them scores of miles and smash them down on faraway houses; it can tear up railroad tracks, topple water towers, rip a scar-like swathe through a hitherto tidy prairie town and smash hospitals and schools and churches with godless abandon. Seldom is a tornadic storm more than a few hundred feet wide. It may also retain its identity and its structural integrity for many miles of horizontal travel, guided by its cumulonimbus puppet master hovering above and from which it was first spawned. It is perhaps invidious to use the word *integrity* in connection with so pitiless a phenomenon, but tornadic integrity is truly what we are talking about—and if such a wind monster can cruise its way uninterrupted across a floodplain or a prairie, the damage and destruction and obliteration it can cause may be legendary. Localized, to be sure—tornadoes are officially classified as non-synoptic windstorms, meaning that they do not affect the huge areas that a large-scale cyclone or a regional weather disturbance might. But within that tightly defined area, the misery caused by such a storm can take generations for its victims to forget.

{ 22 }

I began this book with a description of the events of March 24, 2023, in Rolling Fork, Mississippi, when a tornado near-totally destroyed the town. Seventeen people died as a result of the seventy-one-minute passage of the storm—in a die-straight line from southwest to northeast—through the town and the nearby hamlet of Silver City. The damage was of an extent and a savagery of which locals will doubtless speak for generations. Among the buildings destroyed or badly hurt were the hospital, the high and elementary schools, a gathering of cotton and tobacco warehouses, the library, water tower, animal shelter, police department, post office, city hall—almost all of official Rolling Fork was obliterated, and houses by the score were just wiped out, torn from their foundations, their bricks and tiles scattered like chaff to the horizon. One elderly couple was killed when a semitrailer truck was lifted from half a mile away, thrown into the air, and dumped back down on their roof, crushing them under the wreckage.

I had gone to Rolling Fork six months after the storm, in the company of a woman whose home had been destroyed and whose neighbors had been similarly dispossessed, rendering the calamity somewhat more bearable, if shocking. By then, living many miles away in a rented bungalow, she was phlegmatic about her losses, wanting only to be able to go home, and to have a home to go to. She knew that the Rolling Fork tornado was seen by some meteorologists as a kind of bellwether, an illustration of the gathering notion that Tornado Alley, that area many miles to the west across the Mississippi River where storms have for decades past traditionally proliferated, on the plains of Kansas and Oklahoma and the Texas Panhandle,* seemed now to be shifting eastward, and putting such states as hers in the crosshairs. She knew this, but still she wanted to go home.

The day we visited was hot and sultry, the air thick as felt

---

* So frequent are tornadoes in western Texas that at Texas Tech's National Wind Institute in Lubbock they have built a machine, the VorTECH, that creates artificial ones, and you can crawl into an armored Plexiglas viewing chamber directly below its funnel and watch what happens when a large balk of timber is thrown into its maw to become, if not instantly shredded, then a violently dangerous unguided missile. The vertical tube, a sort of high-rise wind tunnel, can only reproduce the mildest of storms, but students see it as much more educational than a colored diagram of the complex mechanics of rotating air. The institute was set up after the 1970 Lubbock Tornado, which killed twenty-six and did damage estimated at $250 million.

and perfectly still. We watched as a cattle egret soared over a bayou and a catfish tipped his lip above the surface of the water. The town was quite soundless: no air-conditioners growling, no cars honking, no doors slamming, no people chattering. Once in a while a truck whistled past on Highway 61, but otherwise Rolling Fork was just about dead. *We need rain,* remarked my friend, and nodded toward a dark squall cloud far away to the west, on the Louisiana side of the river. *Let's hope the wind will blow that one our way.*

I called her a year later, on Christmas Day. She and her family had finally moved back in, and most of her neighbors had too. The town was slowly coming back to life. She didn't imagine there would be another tornado in Rolling Fork, whatever the changing patterns suggested, so she would chance it. She was not afraid, for sure. We spoke a little of the wind more generally—that though it could destroy, it also brought good things, like the rain shower that she remembered we had wanted brought across from the Louisiana side of the river that hot summer's afternoon the year before. I offered her one morsel of wind trivia—that despite winds being generally invisible, a tornado was, like a waterspout and a prairie dust storm, something that because of all the debris it tore up and carried with it was just about the only kind of wind that was wholly visible. She thought for a moment, harking back to the tornado that had so wrecked her little hometown nearly two years before. *I don't know if it was visible or not,* she replied. *After all, it was night. It was pitch-dark. For those who lived through it it must have been impossible to tell.*

{ 23 }

For now, there is for my Mississippi friend only one wind-borne certainty: that in every succeeding early American summer, there will be more tornadoes, somewhere nearby. They come as regularly as the first cuckoos, or the returning swallows of Capistrano. And the same is true of all winds of all kind—whether northeast trade winds or Roaring Forties gales or howling Drake Passage monster storms; whether Santa Anas or siroccos or boras or harmattans, whether brickfielders or burans, chinooks or Cape Doctors, haboobs or sonoras, kamikazes or levanters, simooms or zephyrs, tramontanas or typhoons; whether a witch, waft, a williwaw, or a zonda. They may be westerlies, northerlies, easterlies, or southerlies, or winds named for any compass point in between. Each may blow as a breeze or a full gale, a cat's-paw or the lightest of airs or a wind that will dry the clothes on a line and leave them freshly scented with the purest of pure air. They will bring those clouds across the river from Louisiana—because that is what winds have the power and mood to do—to bring all the goodness of weather with them, to help keep the world alive.

There is some talk—less now than a decade past; less as I finish this book than when I started it—of a period of Global Stilling, of a wind drought, or a quiescence; and for a while it may well be true that changes in temperature differentials around the planet may cause some easing, some interruption, some disruption to patterns long known and offering comfort through their familiarity.

There are sure to be some consequent changes in the manner in which our atmosphere shifts itself as it drapes around and protects all life beneath. Of all the components of that atmospheric shifting, wind is the undisputed prime. A world without wind is just too dreadful to contemplate. Rather, one trusts and prays, it shall remain as it has been since creation—invisible, eternal, and essential.

WITH GRATITUDE

This was a most enjoyable and absorbing book both to research and to write. Naturally, I hope many will think it a satisfying read. I was generously assisted by multitudes: those whose personal or professional interest in the subject of wind happened to tend more to the romantic than to the mathematic were especially happy to join me in celebrating the story of this invisible and essential magic.

If I had to single out one of these enthusiasts in particular, it would be that heroic eccentric of the northern Adriatic, Rino Lombardi, who for the last twenty years has run his one-room Museo della Bora in a back street of Trieste. All who visit leave enthralled and educated; I remain forever in his debt.

Among others similarly enchanted by wind's many mysteries, and who shared their rapture with me, I would note Rachel Cobb, who devoted many years to photographing the Mistral in Provence; to Jim Robbins, who knows how birds in flight deal with sudden wind gusts, and who from his writing base in Montana first alerted me to the probable existence of the phenomenon of Global Terrestrial Stilling; to Professor Takuji Waseda at the University of Tokyo, who led me to the studies that connected maritime wave heights to wind speed and which famously caused the 1944 D-Day landings being staged on June 6, not June 5, as had been previously planned; to my old friend Mizue Iijima at JAMSTEC, Yokohama, who introduced me to Professor Waseda; to Emma Si Nae of San Francisco, who kindly took a hundred-mile road-trip detour to a near-deserted hamlet in the Oklahoma panhandle to find the first of FDR's wind-calming shelterbelt plantings, and who transcribed the only vaguely legible historical marker there; I have no words sufficient to thank Kristi Cardoso of the Liljestrand Foundation in Hawaii, curating a house whose unique architecture makes full use of the local trade winds to keep it eternally cool and comfortable; to Cesar Azorin-Molina of the Spanish National Research Council in Madrid, an expert in Mediterranean winds; and to Leslie Stephenson, fortunate and brave survivor of a deadly Mississippi tornado.

My now longtime editor at HarperCollins New York, Sara Nelson, more than ably gave this book shape and purpose, and did so with kindness and grace; she was once again helped with great aplomb by the vastly talented Edie Astley, now deservedly a rising star. Across in London the book was nurtured by the much-admired and very dear Arabella Pike. To this holy trinity of publishing talent I offer humble thanks, once more. As I thank my agent, Suzanne Gluck, of the William Morris Agency, who, in concert with her husband, Tom Dyja, to whom I dedicate this book—please *qv* the page—championed the idea from its very beginnings.

My oldest son, Rupert, from his peaceful Somerset village, performed a congeries of research tasks for me, as he has for earlier books over the last three decades. My gratitude knows no bounds. Likewise my most profound thanks to his younger brother, Angus, who drove me four thousand cross-country miles to and from Lubbock in the west Texas panhandle, there to meet the amiably helpful Anna Thomas, director of the National Wind Institute, who among her many toys runs the previously mentioned machine that generates homemade tornadoes. And finally I am grateful, too, of course, for the help and advice offered by my wife, Setsuko, who put up with all the absences and silences for which a writer's life is known, and that demand a measure of long-suffering patience, which in this family was happily offered and gratefully accepted.

Simon Winchester
Sandisfield, Massachusetts, August 2025

GLOSSARY

This list, edited from a catalog famously assembled half a century ago by Lyall Watson, is roughly limited to names of winds that blow *today*, and omits all but one of the multitude from classical times. In most cases I list the basic meteorological attributes of each—direction, temperature, associated weather, as with "NE; cold; southern France") and add material of interest for the better known or more unusual.

Arashi: N; cold, occasional fog; Japan.

Arifi: S; hot and dry; Morocco.

Bad-i-sad-o-bist-roz: N; hot; Iran and Afghanistan.

Barat: NW; hot; Sulawesi, Indonesia.

Barber: N; bitter cold; Canada.

Barine: W; onshore gale; Venezuela.

Bayomo: N; violent; Cuba.

Belat: N; cool; Oman and Saudi Arabia.

Berg: N; hot; South Africa.

Bhoot: Variable; dust devil; India.

Bise: NE; cold; Languedoc.

Boekifu: NE; cool; Japan.

Bofu: Gale; Japan.

Bohorok: Warm; Sumatra.

Bora: NE; sustained, violent, frigid; Trieste.

Borasco: Gusty; Mediterranean.

Bornan: N; cold; Geneva.
Breva: N; cool; Lake Como.
Brickfielder: SW; hot, dry; eastern Australia.
Brisa: NE; cool, moist; Philippines.
Bruscha: NW; cold; Switzerland.
Buran: NE; cold; Russia.
Burga: NE; cold, snow-laden; Mongolia, Alaska.
Buster: S; cold, wet; Australia.
Cacimbo: SW; cool; Angola.
Cantalaise: N; cold, snowy; France.
Cape Doctor: SE; cool, blows smog away from Cape Town.
Cat's Paw: Faint breeze; United States.
Challiho: S; pPrecursor of SW monsoon; eastern India.
Chergu: S; warm; Morocco.
Chi'ing fung: Faint breeze, China.
Chili: S; hot, dry; Morocco, Tunisia, Algeria.
Chinook: W; warm, katabatic; western United States.
Chocolatero: Hot, sandy dust-colored; Mexico.
Chom: S; hot, dry; Algeria.
Chubasco: Violent gale, Gulf of California.
Cierco: W; cold; Spain.
Collada: N; cold; Gulf of California.
Coromell: Breeze; Mexico.
Coronazo: SE; Pacific coast of Mexico
Criador: W; rainy; northern Spain.
Crivetz: NE; cold; Romania.
Demani: Breeze; Uganda.
Drinet: Cold katabatic; Romania.
Duster: W; laden with soil; plains in the United States.
Easter: E; strong; Oregon.
Elephanta: S; gale; India, Malabar coast.
Elvegast: E; cold, dry; Norway.
Erh chi chih fung: N; cold; China, Mongolia.

Etesian: NW; cool; Greece.

Fakatiu: NW; cool; Melanesia.

Feh: Breeze; Shanghai.

Flakt: Gentle breeze; Sweden

Flauwe: Gentle breeze; Holland.

Fohn: Hot, dry; katabatic; Switzerland.

Frisk: Gale; Sweden.

Fuga: Strong gale; Crimea.

Galerna: NW; cold, squally; Bay of Biscay.

Garbin: SW; moist; western Spain.

Garmsal: W; hot, dusty; Turkestan.

Garvi: S; Algeria.

Gending: Strong; Java.

Geneva: SW; rainy; Switzerland.

Gergui: S; hot, dry; sandstorms; Algeria.

Gerona: Hot, dry; Spain.

Gharbi: SW; hot, moist, Moroccan dust-laden with red-rain; France, Italy, Greece.

Ghibli: S; hot desert; Tunisia.

Giba: Strong "hang-horse" gale; Japan.

Gregale: NE; strong; Greece, from Balkans.

Grenoble: SW; rainy; southern France.

Haboob: Dust storm, preceded by 3,000-foot-tall wall of dust, then rain; Sudan.

Halne: Katabatic gale, Czech Republic.

Harmattan: NE; dust-laden, West African coast.

Haur: E; North Africa.

Hawajanubi: S; Arabian Peninsula.

Hawashimali: N; Arabian Peninsula.

Hayate: Gale; Japan.

Helm: NE; Pennine Hills, England.

Hokuto: NE; Japan.

Inverna: Downdraft; Lake Maggiore.

Iseran: N; cold, gusty; French Alps.
Jura: Cold, gusty; Jura foothills.
Kabeyun: W; cold; northeastern United States.
Kamakaza: N; cold; Japan.
Kamikaze: "Spirit wind"; Japan.
Kapalilua: Sea breeze; Hawaii.
Karaburan: ENE; "black wind" of the Gobi; China.
Kawaihae: Squall; Hawaii.
Khamsin: S; hot, dusty; Egypt ("the ninth plague of Egypt").
Kibibonokka: N; cold; United States.
Knik: SE; strong; Alaska.
Kochi: E; Japan.
Kogarashi: N; cold winter wind.
Kohala: Gale; Hawaii.
Kokaze: "Little wind"; Japan.
Kolawaik: S; cool; Argentina.
Kona: SE; sultry; Hawaii.
Koochee: Whirlwind; Australia.
Koshava: NE; cold, snow-laden; Balkans.
L'Este: SE; hot, dry; Madeira.
Laawan: W; grain-winnowing; Morocco ("the helper").
Labech: SW; rainy; Provence.
Lakawa: E; Argentina.
Laxwaik: W; central Argentina.
Levanter: E; Balearic Islands.
Libeccio: SW; sea breeze; Italy.
Liberator: W; moist; Gibraltar.
Ljuka: E; warm, katabatic; Balkans.
Lombarde: Warm, alpine; France.
Maestral: N; cold, strong; Gulf of Genoa.
Maestro: NW; summertime; western Adriatic.
Maledetto: N; Alpine; Italy.
Maloja: Alpine Switzerland.

Mamatelel: NW; hot; Malta.

Maoifung: NE; China.

Marin: SE; hot, heavy rain; southern France.

Matsubori: SE; crop-damaging; northern Kyushu, Japan.

Matsukaze: Light "pine-rustling" breeze; Japan.

Mbatis: S; light evening breeze; Greece.

Melamboreas: N; similar to Mistral; Provence.

Melteme: NE; autumnal evening; Greece and Turkey.

Mezzer-Ifoullousen: SE; violent, cold; Morocco.

Minuano: SW; cold; southern Brazil and Uruguay.

Mistral: NW; cold, blustery, mood-altering wind of Alpine origin, much affecting Provence.

Monçao: NE; Portugal.

Monsoon: NW or SW, seasonal, bringing much-needed summer rains; India.

Morget: Nighttime land breeze; Lake Geneva.

Mracna Bura: NE; cold; Croatia.

Myatel: NE; violent; northern Russia.

Naalehu: Dry, land breeze; Hawaii.

Narai: NE; very cold Siberian wind; in Japan.

Nasim: Breeze; Saudi Arabia.

Nekrayak: NE; snow-laden; Greenland.

Nor'easter: NE; cold seasonal wind; coastal New England.

North Canadian: N; Quebec.

Norther: N; unusual burst of arctic air in Texas.

Nor'wester: NW; katabatic; Christchurch, New Zealand.

Nowaki: Hot, autumnal; Japan.

OE: Sudden rotating storm; Faroe Islands.

Oes: Waterspout; Netherlands.

Ora: S; morning breeze; Lake Garda, Italy.

Orkan: Gale; Norway.

Orleans: E; Loire Valley, France

Oroshi: NE; cold, dry; Kanto Plain, Japan.

Ouari: Summertime sandstorm; Somalia.
Pampero: SW; cold, rainy; central Argentina.
Papagayos: N; pleasantly cool; Costa Rica.
Pittarak: NW; dry; Greenland.
Polack: NE; cold, dry; Sudetenland.
Ponente: W; sea breeze; Italian west coast.
Pontia: N; cold, dry; from Rhone Valley.
Poriaz: NE; summertime cold from Black Sea, over Bulgaria.
Prodromes: NE; mild, precedes rising of Sirius over classical Greece. Until recently, Prodrome was used as the telegraphic address of most British diplomatic missions, which were charged with anticipating faraway events of possible interest to London.
Pruga: N; strong, cold; Alaska.
Puelche: E; katabatic Andean wind; Chile.
Purga: NE; strong, snow-laden; Russia.
Raghieh: E; cold; coastal Syria.
Raiklas: NE; cold; coastal Turkey.
Raki: W; Melanesian Islands.
Rebat: S; morning breeze; Lake Geneva.
Reppu: Circular, tropical; Japan.
Reshabar: NE; dry, dusty; Kurdistan.
Roeteturm: N; katabatic; Romania.
Rok: Gale; Iceland.
Samiel: Hot, dry; Turkey.
Sansar: NW; icy; Iran.
Santa Ana: NE; strong, evil reputation; California.
Seguin: Breeze; Provence.
Seistan: NE; cold, strong; Iran.
Shamal: NW; mild, summertime; Mesopotamia.
Shamsir: N; cold; Iran.
Sharav: Hot, dry, desert; Israel.
Sharkiye: Cool, desert; Lebanon.
Shih lung: NE; China.

Shimpu: "Divine" wind of western Japan.

Siffanto: Hot, heel of Italy.

Simoom: Hot, dry; North Africa.

Sirocco: SE; hot, springtime, blowing off the Sahara.

Sno: E; cold; Norway.

Solano: SE; dusty; sea breeze; Spain.

Sonora: Summer, desert; Arizona.

Souledre: NE; cold; France.

Soyo kaze: Gentle breeze; Japan.

Steppenwind: NE; cold; Germany.

Stikine: N; gusty, strong; Alaska.

Suestada: N; rain-bearing; Argentina, Uruguay.

Suhaili: SW; cold, wet, strong; coast of Iran.

Sukhovey: E; warm, dusty; Gobi Desert, China.

Surazo: E; cold, strong; Peru.

Sveszhest: Gentle breeze; Russia.

SZ: Gentle breeze; China.

Taku: NE; strong; Alaska.

Tanga mbili: Variable, in September; East Africa.

Tapayagua: Squall; Central America.

Tarai: SW; monsoon; Andaman Islands.

Tatsumaki: Tornado; "dragon whirl"; Japan.

Tebbad: Hot, dusty; Turkestan.

Tegen: NE; Holland.

Tehuantepecer: N; strong, cold; Pacific Mexico.

Terral: Land breeze; Chile.

Tezcatlipoca: N; divine wind of the Aztecs.

Thalwind: Valley breeze; Germany.

Thar: Hot, dry; Rajasthan.

Tivano: S; midday breeze; Lake Como.

Tokalau: NE; Fiji.

Tramontana: S; cold, strong, Alpine; Lake Maggiore.

Trauben-Kocher: Warm; Fohn type; "grape cooker"; Switzerland.

Tsuji: Wind of the crossroads"; Japan.

Tsumujikaze: Whirlwind; Japan.

Tuaura: S; Melanesia.

Tung shan: NE; trade; China.

Uberre: Warm, katabatic; Neuchâtel.

Vallesaria: Fohn type. East of Lake Geneva.

Vardarac: NW; cold, dry; Balkans.

Vendavales: SW; winter; hot, gentle, then rain; Morocco.

Vento coado: Gentle, mountain; Portugal.

Vento de baixo: Sea breeze; Portugal.

Vind-blaer: Breeze; Iceland.

Virazon: Sea breeze; coastal Chile and Spain.

Viuga: NE; cold, stormy; Siberia.

Vyetorok: Gentle breeze; Russia.

Wabuan: E; "morning bringer" of the Algonquins; United States.

Waddy: Puff of wind; Cornwall.

Waimea: Misty sea breeze; Hawaii.

Warm braw: SW; warm, dry; New Guinea.

Waryaraik: N; Gran Chaco, Argentina.

Watakushi: The "me" wind of Japan.

Whirly: Violent, short-lived storm; Antarctica.

Whittle: Gusty; Cheshire, England.

Williwaw: Brief, blustery wind; Strait of Magellan.

Willy-willy: Circular storm; Timor Sea.

Wisper: Cool, evening bluster in Rhine Valley, Germany.

Witch: Another name for Santa Ana winds.

Xlokk: Hot, dry; Malta.

Yama oroshi: Katabatic; Japan.

Yamo: Whirlwind; Uganda.

Zephyr: Pleasantly warm breeze; Italy.

Zonda: W; hot, dry, katabatic; from Argentinian Andes.

BOOKS I CONSULTED AND RECOMMENDED FURTHER READING

Azorin-Molina, Cesar, et al. "Homogenization and Assessment of Observed Near-Surface Wind Speed Trends over Spain and Portugal, 1961–2011." American Meteorological Society. *Journal of Climate* (2014).

Bagnold, R. A. *The Physics of Blown Sand and Desert Dunes.* Methuen, 1954.

Bathurst, Bella. *The Lighthouse Stevensons.* Harper, 1999.

Baum, L. Frank. *The Wonderful Wizard of Oz.* George M. Hill Company, 1900.

Belci, Corrado. *Il Libro Della Bora.* Edizione Lint, 2003.

Brown, Slater. *World of the Wind.* Alvin Redman, 1961.

Cobb, Rachel. *Mistral: The Legendary Wind of Provence.* Damiani, 2018.

Conrad, Joseph. *Typhoon.* G. P. Putnam's Sons, 1902.

Corbin, Alain. *A History of the Wind.* Polity, 2023.

Daston, Lorraine, and Elizabeth Lunbeck, eds. *Histories of Scientific Observation.* University of Chicago Press, 2011.

DeBlieu, Jan. *Wind: How the Flow of Air Has Shaped Life, Myth, and the Land.* Houghton Mifflin, 1998.

Dee, Tim. *The Running Sky: A Birdwatching Life.* Cape, 2009.

Defoe, Daniel. *The Storm.* John Nutt, 1704.

DeHarpporte, Dean. *Wind Atlas of the United States.* 3 vols. Van Nostrand Reinhold, 1983.

DeVilliers, Marq. *Windswept: The Story of Wind and Weather.* Walker, 2006.

Dolin, Eric Jay. *A Furious Sky: The Five-Hundred Year History of America's Hurricanes.* Liveright, 2020.

Droze, Wilmon Henry. *Trees, Prairies, and People: A History of Tree Planting in the Plains States.* Texas Woman's University, 1977.

Dutton, John A. *The Ceaseless Wind: An Introduction to the Theory of Atmospheric Motion.* McGraw-Hill, 1976.

Egan, Timothy. *The Worst Hard Time: The Untold Story of Those Who Survived the Great American Dust Bowl.* Mariner Books, 2006.

Emanuel, Kerry. *Divine Wind: The History and Science of Hurricanes.* Oxford University Press, 2005.

Ferrel, William. *A Popular Treatise on the Winds.* Wiley, 1890.

Freeman, John. *Wind, Trees.* Copper Canyon Press, 2022.

Friendly, Alfred. *Beaufort of the Admiralty.* Random House, 1977.

Glickman, Todd, ed. *Glossary of Meteorology.* 2nd ed. American Meteorological Society, 2000.

Grundy, Capt. Josiah, and Wilkins W. Wheatly. *Square Riggers Before the Wind.* Dutton, 1939.

Hangan, Horia, and Ahsan Kareem. *The Oxford Handbook of Non-Synoptic Wind Storms.* Oxford University Press, 2021.

Herschel, Sir John. *A Manual of Scientific Enquiry Prepared for the Use of Officers in Her Majesty's Navy.* The Admiralty, 1886.

Hidy, George M. *The Winds: The Origins and Behavior of Atmospheric Motion.* Van Nostrand, 1967.

Hughes, David McDermott. *Who Owns the Wind?* Verso, 2021.

Hughes, Richard. *In Hazard.* Chatto & Windus. London, 1938.

Huler, Scott. *Defining the Wind: The Beaufort Scale, and How a Nineteenth Century Admiral Turned Science into Poetry.* Crown, 2004.

Hunt, Nick. *Where the Wild Winds Are.* Nicholas Brealey (John Murray), 2017.

Huntington, Ellsworth. *Civilization and Climate.* Yale University Press, 1924.

Huschke, Ralph E. *Glossary of Meteorology.* American Meteorological Society, 1970.

Hydrographer of the Navy. *Ocean Passages for the World.* UK Ministry of Defence, 1973.

Inwards, Richard. *Weather Lore.* Rider and Company, 1950.

Johnson, Captain Irving. *Around Cape Horn.* Mystic Seaport, 1980 (video).

Kamkwamba, William. *The Boy Who Harnessed the Wind.* HarperCollins, 2009.

Lindau, Ralf. *Climate Atlas of the Atlantic Ocean.* Springer, 2001.

Linden, Eugene. *The Winds of Change.* Simon & Schuster, 2006.

Lords Commissioners of the Admiralty. *A Seaman's Pocket-Book.* London, 1943.

Maury, M. F. *The Physical Geography of the Sea and its Meteorology.* Harper & Brothers, 1861.

McNeley, James Kale. *Holy Wind in Navajo Philosophy.* University of Arizona Press, 1981.

Melville, Herman. *Moby-Dick; or, The Whale.* Harper & Brothers, 1851.

Monmonier, Mark. *Air Apparent.* University of Chicago Press, 1999.

Morris, Jan. *Trieste and the Meaning of Nowhere.* DaCapo Press, 2001.

Needham, Joseph. *Science and Civilisation in China.* Vol. 4, part 3. Cambridge University Press, 1971.

Neer, Robert. *Napalm: An American Biography.* Belknap (Harvard University Press), 2013.

O'Brien, Gregory, and Louise White. *Big Weather: Poems of Wellington.* Mallinson Rendel, 2000.

Pelletier, Cathie. *Northeaster.* Pegasus Books, 2023.

Price, Trevor J. *James Blyth: Britain's First Modern Wind Power Pioneer.* Sage Journals/*Wind Engineering* (2005).

Raban, Jonathan, ed. *The Oxford Book of the Sea.* Oxford University Press, 1992.

Ramage, C. S. *Monsoon Meteorology.* Academic Press, 1971.

Rappaport, Captain Elliot. *Reading the Glass.* Hodder, 2023.

Redniss, Lauren. *Thunder and Lightning: Weather Past, Present, Future.* Random House, 2015.

Sakamoto, Dean, ed. *Hawaiian Modern: The Architecture of Vladimir Ossipoff.* Yale University Press, 2015.

Schwartz, Stuart B. *Sea of Storms: A History of Hurricanes in the Greater Caribbean from Columbus to Katrina.* Princeton University Press, 2015.

Seed, Patricia. *Sails and Shadows: How the Portuguese Opened the Atlantic and Launched the Slave Trade.* University of California Press, 2026.

Stegner, Wallace. *Angle of Repose.* Doubleday, 1971.

Steinbeck, John. *The Grapes of Wrath.* Viking, 1939.

Streever, Bill. *And Soon I Heard a Roaring Wind: A Natural History of Moving Air.* Little, Brown, 2016.

Svenvold, Mark. *Big Weather: Chasing Tornadoes in the Heart of America.* Henry Holt, 2005.

Tufano, Fabio. *Antropologia della Bora.* Museo della Bora, 2023.

Vanhoenacker, Mark. *Skyfaring: A Journey with a Pilot.* Alfred A. Knopf, 2015.

Wailes, Rex. *The English Windmill.* Routledge & Kegan Paul, 1954.

Walker, Gabrielle. *An Ocean of Air.* Bloomsbury, 2007.

Watson, Lyall. *Heaven's Breath. A Natural History of the Wind.* William Morrow, 1984.

Weiss, Allen S. *The Wind and the Source: In the Shadow of Mont Ventoux.* SUNY Press, 2005.

Yvart, Jacques. *The Rising of the Wind: Adventures Along the Beaufort Scale.* Green Tiger Press, 1983.

Zafón, Carlos Ruiz. *The Shadow of the Wind.* Penguin, 2001.

INDEX

Italicized page numbers refer to photographs and their captions. Page numbers followed by n indicate notes.